The Future of the Citizen-Soldier Force

The Future of the Citizen-Soldier Force

Issues and Answers

JEFFREY A. JACOBS

THE UNIVERSITY PRESS OF KENTUCKY

Copyright © 1994 by The University Press of Kentucky

Scholarly publisher for the Commonwealth, serving Bellarmine College, Berea College, Centre College of Kentucky, Eastern Kentucky University, The Filson Club, Georgetown College, Kentucky Historical Society, Kentucky State University, Morehead State University, Murray State University, Northern Kentucky University, Transylvania University, University of Kentucky, University of Louisville, and Western Kentucky University.

Editorial and Sales Offices: Lexington, Kentucky 40508-4008

Library of Congress Cataloging-in-Publication Data

Jacobs, Jeffrey A., 1957-
    The future of the citizen-soldier force : issues and answers / Jeffrey A. Jacobs.
      p.   cm.
    Includes bibliographical references and index.
    ISBN 978-0-8131-5621-7
    1. United States. Army reserve. I. Title.
UA42.J33   1993
355.3'7'0973—dc20   93-28304

To the memory of
Colonel Douglas M. Craver, United States Army

"Once a paratrooper, always a paratrooper"

CONTENTS

List of Figures and Tables	ix
Acknowledgments	xi
1. The Three-Army System	1
2. The Evolution of Three Armies	26
3. Systemic Disconnects in the Total Army Circuit	50
4. Geography, Time, and Other Readiness Detractors	68
5. Lessons Learned from Desert Storm	88
6. Realizing the Potential of the Reserve Components	104
7. Through the Political Minefield	136
Appendix	147
Glossary	159
Notes	163
Selected Bibliography	175
Index	181

FIGURES & TABLES

Figures

1. Total Army Structure — 6
2. Structure of the Army's Reserve Components — 8
3. Lines of Authority for the Army's Reserve Components — 14
4. Hypothetical CAPSTONE Alignment — 17
5. Proposed Organization for U.S. Army Reserve or Fully Federal Reserve Force — 116
6. Department of Defense Structure — 150

Tables

1. Classification of Army Units — 12
2. The CONUSAs and Theater Armies — 15
3. Mobilization Categories — 20
4. Army Training Readiness Levels — 91
5. The Combatant Commands and Their Army Components — 148
6. Major Units of the Army Force Structure — 152
7. Selected Information Concerning Army Leaders and Commanders — 154
8. The Reserve Components' Contribution to the Total Army — 156

ACKNOWLEDGMENTS

THIS book would not have been possible without the help of several people. I gratefully acknowledge the assistance of my father-in-law, Major General Robert B. Solomon, U.S. Army Retired, for his critical analysis and suggestions. I am also grateful to my father, Colonel Irwin M. Jacobs, U.S. Army Retired, for his editorial assistance. Professor Raymond A. Moore also assisted me greatly with his advice and incisive comments. Finally, I thank my family, Sharon, Sammy, and Julie, for lovingly, if not always willingly, affording me the time to tilt at windmills.

1. THE THREE-ARMY SYSTEM

> If American citizen armies, extemporized after the outbreak of war, could do as well as Washington's Continentals and as well as the citizen armies of Grant and Lee, what might they not do if organized and trained in time of peace?
> —John McAuley Palmer

THE defense policy of the United States is at a crucial crossroads. The Soviet Union, the threat for which the military has trained since World War II, has imploded. Domestic budgetary constraints, although they have always been an essential consideration in the formulation of defense policy, recently have assumed even greater importance. Given the changed threat and the reality of less money, defense policy makers face significant decisions concerning many issues—roles, missions, force structure, and procurement, to name but a few.

This book is about one of those issues: the future shape of the reserve components of the United States Army. Although B-2 bombers, M-1 tanks, and nuclear submarines may command larger headlines, the reserve issue is an important one that cannot be shunted aside as the course of the military is mapped out in Congress and the Pentagon.

Historically, the active Army has been a self-contained force that has required augmentation only in the event of a large-scale conflict requiring national mobilization, on the order of World Wars I and II. Until relatively recently, the role of the Army's reserve components was to provide "forces in reserve"—a source of available manpower to supplement the active Army after it had reached the limit of its capabilities. The reserves generally were not ready to fight when called, and in some instances, volunteers and draftees played an equal or greater wartime role in expanding the Army.

Although the Vietnam War exceeded the capabilities of the peacetime active Army, the Army did not mobilize the reserves for that conflict but instead expanded with draftees. Dissatisfied with both the military and the political ramifications of the decision not to mobilize the reserves, the Department of Defense implemented the Total Force Policy in 1973. That policy, which governs the nation's military reserves today, changed the role of the reserve components drastically.

The objective of the Total Force Policy is to integrate the services' reserve components with their active components, obviating the necessity to resort to the draft. Under the policy, the Department of Defense has deliberately placed significant responsibility for the nation's defense on the shoulders of the citizen-soldier; the reserves, rather than draftees, are now the nation's primary source of available military manpower when expansion of the military beyond the size of the active force is required. The reserves, therefore, are no longer "forces in reserve" but a key cog in the United States' military apparatus. In fact, the deployment of the reserves to the Persian Gulf for Operation Desert Shield/Desert Storm was absolutely essential, because active forces could not have done the job alone.

As Desert Storm proved, the benefits of the Total Force Policy have been many. From the Total Army perspective, trained reserve units have provided the unit continuity so lacking during Vietnam. The policy also has put a strong military presence back into many civilian communities and has helped ensure that the employment of sizable American military forces will receive widespread political support. From the perspective of the Army's reserve components, the Total Force Policy has infused the Army National Guard and the Army Reserve with money for equipment, training, facilities, and personnel.

Desert Storm, however, exposed some weaknesses in the Total Force Policy. And although Desert Storm answered many questions about the policy, it left many questions unanswered, and it raised many others. Does the policy work as advertised, or is the concept at least partially flawed?

The Total Force Policy has not lived up to expectations in many respects. The fundamental structure of the reserves has changed little since the 1920s.[1] Superimposing the Total Force Policy on that structure in effect put a 1973 engine on a 1920 chassis. Moreover, because the Army as an institution has tended toward gradual evolution rather than more rapid change, the inertia of tradition has been difficult to overcome. The influence of this tradition, coupled with the fact that the Guard and Reserve have always been powerful political organizations (the Guard more so than the Reserve), has impeded a smooth transition from the rhetoric of the Total Force Policy into the reality of an effective Total Army, one in which all three of the Army's components—the active Army, the Army National Guard, and the Army Reserve—are truly integrated.

The past few years have been ones of dramatic world change, and the world continues to evolve in ways that were inconceivable only a decade ago. As the threat changes and as the dollars dwindle, defense policy makers are examining the Army from top to bottom, and, consequently, the Army is changing significantly.

As major portions of the Army, the Army National Guard and the Army Reserve will not escape scrutiny. Indeed, change in the Army's reserve components is inevitable. Just how the reserves will change, however, is a subject of some controversy. What should be the role of the reserves in the "new" Army? Should the Total Force Policy be retained as is, modified, or scrapped? What are the implications of Desert Storm vis-à-vis the reserves? If the reserves are forces unto themselves, if they do not fit smoothly into the wartime structure of the active Army, and if they are not as ready as they are expected to be, are they military liabilities in the context of a conflict on a scale broader than Desert Storm?

To those not familiar with the role of the armed forces in national policy, these questions may not seem to be of overriding national import. But they are. Historically, despite their best political efforts, the reserves have remained in the background of the public's—and the Army's—consciousness until the shooting starts. But as the Army inevitably shrinks, the importance of the citizen-soldier force will be magnified.

Although the Soviet threat may have disappeared, the United States will not become an isolationist nation, as it did after World War I. The world today is neither stable nor peaceful: witness the breakup of Yugoslavia, Middle East tensions, civil unrest in the Caribbean, and the proliferation of nuclear and chemical weapons. Even without the specter of the Soviets, threats to the United States' interests persist throughout the world, as Desert Storm bore out. As General Colin L. Powell, the chairman of the Joint

Chiefs of Staff, has written, "The real threat we now face is the threat of the unknown, the uncertain. The threat is instability and being unprepared to handle a crisis or war that no one predicted or expected."[2] Moreover, although the commonwealth that has replaced the Soviet Union is certainly not a foreseeable threat, given the current instability of that commonwealth and the fact that the former Soviet republics collectively still possess a significant nuclear capability and one of the largest armies in the world, the new commonwealth's capabilities cannot be ignored completely.

As Desert Storm also proved, the active Army is not totally self-sufficient even against foes less formidable than the former Soviet Union. Indeed, the Army is now four active divisions smaller than it was in 1990 (and in 1995 it will be six active divisions smaller) and today could not fight Desert Storm without relying even more heavily on the reserves. In fact, the Army probably could not put a force of the same size and capability on the battlefield today without employing reserve infantry and armor units, which it did not use in Desert Storm. Therefore, as defense funding levels continue to decline and as long as the threat is that of the unknown, the post-Cold War active Army will be stretched to the limit, and the reserves may have to pick up even more of the slack.

The reserves' increased importance thrusts the Total Force Policy directly into the national spotlight: the failure of the policy to achieve integration in fact as well as in theory could have significant implications on tomorrow's battlefield. If the active Army cannot fight even brush-fire wars without the reserve components, the roles and capabilities of the reserves are indeed national defense policy issues of the first order. To ignore these issues is folly.

The tough questions, therefore, can no longer be ignored. Serious analysts of reserve issues have quietly asked most of these questions for some time. Policy makers, however, have swept them under the rug because of their political volatility. The nation deserves an informed debate on questions of such fundamental importance; at least then the risks of adopting (or continuing) policies that are less than ideal will be apparent and appreciated. The time has come to air these issues in the political arena.

THE THREE-ARMY SYSTEM

To set the stage, an overview of the reserve components' structure is in order (the Appendix contains more information for those not intimately familiar with the workings of the military or the Army). In practice if not policy, the Army is not one integrated service but three separate entities: the active Army, the Army National Guard, and the Army Reserve. Each of these components maintains its own separate bureaucracy, each is funded separately, and each has its own parochial interests.

To perform its mission, the Army is divided into one active component (the Regular Army) and two reserve components—the Army National Guard of the United States (an identity legally distinct from the Army National Guard, as explained later) and the United States Army Reserve. A *reservist* is a member of the reserve components, either the Army Reserve or the Army National Guard; a *Reservist* is a member of the U.S. Army Reserve. Although there are obvious similarities, each of these components is distinctive.

As of this writing, the combat power of the active component is concentrated in fourteen combat divisions.[3] There are also forces independent of these fourteen divisions in the active component, ranging from armored cavalry regiments to corps support units to higher echelon combat service support units and, of course, the peacetime infrastructure necessary to administer and support such a force. (This is, of course, an oversimplification, but the purpose of this description is only to describe the active component sufficiently to understand its relation to the reserve components.)

Of the fourteen active divisions, only six consist wholly of subordinate active component units (although two more have a full active component complement and an "extra" reserve component brigade). Of those six, two are *forward deployed,* or stationed outside the United States, in Germany. The other four wholly active component divisions are contingency, or rapid-deployment, units—the 7th and 25th Infantry Divisions (Light), the 82d Airborne Division, and the 101st Airborne Division (Air Assault).

The active component divisions are subordinate to one of four corps headquarters, which in turn are subordinate to one of the Army's major commands. Units in Europe, for example, are commanded by U.S. Army Europe, those in the Pacific by U.S.

Combat

Infantry
Armor
Artillery
Special Forces
Engineer (Combat)
Aviation (Combat)

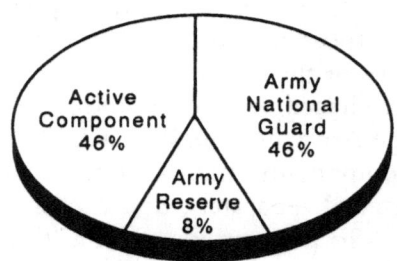

Combat Support

Signal
Chemical
Military Police
Intelligence
Civil Affairs
Engineer
Psychological Operations
Aviation

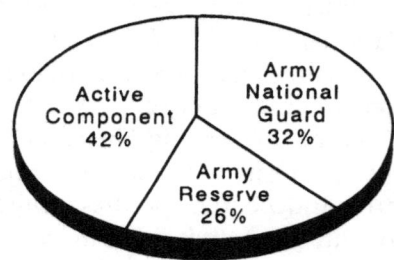

Combat Service Support

Supply and Service
Quartermaster
Transportation
Finance
Ammunition
Judge Advocate
Administration
Petroleum/Water
Railroad
Maintenance
Medical

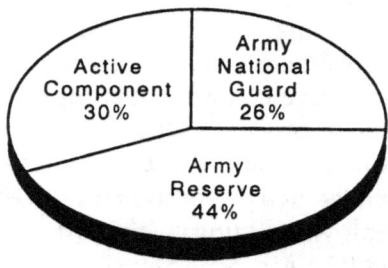

Source: 1990 Reserve Forces Policy Board Report.

Figure 1. Total Army Structure

Army Pacific. The corps and divisions in the continental United States are subordinate to U.S. Forces Command. The commander of Forces Command wears two hats, commanding most of the Army's tactical forces in the continental United States as well as serving as the commander in chief of a specified command with the mission of defending the continental United States.[4]

Although the average citizen pictures only the active component when the Army is mentioned, the active component is, roughly, slightly less than half of the total Army. Under the Total Force Policy, the reserve components are now largely or wholly responsible for many important Army missions. As Operation Desert Storm proved, the Army, at least as it is configured in 1993, cannot perform its mission with active-duty soldiers alone (see fig. 1). Virtually every conceivable operational deployment of any significance will require the mobilization of at least some reserve component units and/or personnel. In fact, even actions as limited as the Grenada operation in 1983 and Operation Just Cause, the invasion of Panama in 1989, required the mobilization of reserve volunteers, despite the characterization of those operations as active component affairs. Indeed, by not mobilizing reserve units involuntarily for Just Cause, the Department of Defense contravened the wishes of the commander in chief of the U.S. Southern Command.

The reserve components are statutorily divided into three categories: the Ready Reserve, the Standby Reserve, and the Retired Reserve (see fig. 2). The Ready Reserve is the most readily mobilized and consists of reserve units and individual reservists not assigned to units. The Ready Reserve is further divided into the Selected Reserve, the Individual Ready Reserve, and the Inactive National Guard. The Selected Reserve consists of reserve units; full-time support personnel (discussed in greater detail below); and individual mobilization augmentees, who are individual Army Reservists assigned to augment active Army organizations, the Selective Service System, or the Federal Emergency Management Agency.

Soldiers in the Selected Reserve train and are paid regularly. Selected Reserve units generally consist of soldiers who live in the same vicinity. These units "drill" one weekend per month and participate in an annual training period, generally fifteen days in length. Individual mobilization augmentees, although they are

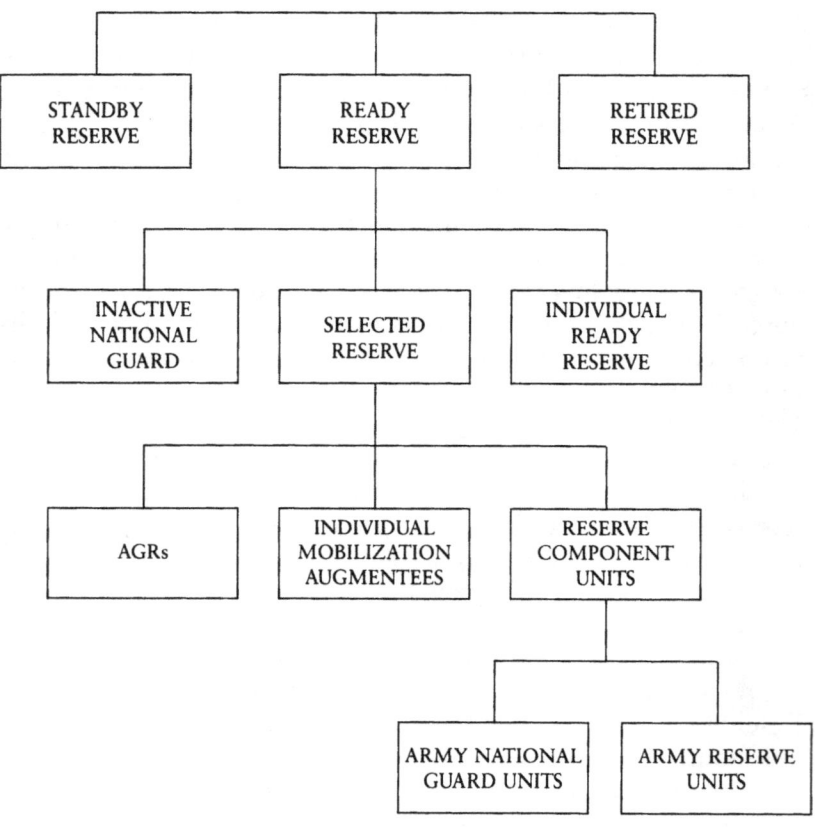

Figure 2. Structure of the Army's Reserve Components

members of the Selected Reserve, are individual members of active component organizations, and most do not participate in monthly drills. Most individual mobilization augmentee positions are found in nontactical organizations (i.e., not in warfighting units), such as higher-level staffs and installation headquarters. The Selected Reserve can be mobilized absent a declaration of war or national emergency; the remainder of the Ready Reserve cannot.

The Individual Ready Reserve is a manpower pool consisting of Reservists who have prior experience in either the active component or the Selected Reserve. Initial enlistment contracts for active Army soldiers require those soldiers to serve a total of eight years—two, three, or four years in the active component and the remainder in the reserve components upon their release from active duty. The Individual Ready Reserve consists of soldiers who have remaining contractual service obligations or who have remained in the Individual Ready Reserve voluntarily upon expiration of their service obligations. These soldiers would be mobilized to bring understrength units to full strength and to replace combat casualties.

The National Guard, which consists of the Army National Guard and the Air National Guard (a reserve component of the Air Force), is the lineal descendant of the colonial militia. Constitutionally, the National Guard continues to be governed both under the militia clauses of the United States Constitution, which empower Congress to prescribe the discipline for the militia, and under the clause that authorizes Congress to raise and support armies.

The Guard is thus a hybrid organization, part state and part federal. Every Army National Guard unit today has two identities. As part of the states' National Guard, the Army National Guard is responsible to the various state and territorial governors, who typically employ the Army National Guard for missions such as civil disturbances and natural disasters. As the Army National Guard of the United States, the Army National Guard is a reserve component of the Army and as such is governed by the Army. The rub is that the Army National Guard must be "federalized" to become the Army National Guard of the United States, and generally federalization occurs only in the case of mobilization (although Guard units can be, and are, federalized for other purposes, such as overseas training).

In peacetime, therefore, the Army National Guard is a state entity, or, more accurately, fifty-four separate entities (each state, the District of Columbia, Puerto Rico, the Virgin Islands, and Guam have a National Guard). The governors command these state and territorial organizations through state or territorial adjutants general or, in the case of the District of Columbia, a commanding general. The adjutants general are major (two-star) generals appointed by the state and territorial governors (South Carolina's adjutant general is popularly elected). The states determine the qualifications for their adjutants general, although, like other National Guard officers, they must meet certain criteria to be "federally recognized" as members of the reserve components of the Army (as state officials, however, federal recognition is not a prerequisite for appointment as an adjutant general).[5] In practice, virtually all adjutants general have risen from within their state Guard ranks, but such progression is not necessarily a requirement. Many adjutants general have prior active service and distinguished combat records.

The states are responsible for mobilizing their Army National Guard units in response to a federal call-up. A state area command is organized in each state to command and control Army National Guard units from the time they are mobilized until they report to their mobilization stations.

Although Army National Guard units are directly responsive to their state or territorial headquarters, tactical Guard units transcend state boundaries. Currently, for example, there are ten combat divisions in the Army National Guard (some of which are slated for inactivation). Eight of these Guard divisions comprise subordinate units from more than one state. Thus, although the headquarters of the 29th Infantry Division (Light) is a part of the Virginia Army National Guard, the division's 3d Brigade, its combat aviation brigade headquarters, and its division support command headquarters all belong to the Maryland Army National Guard. As but one example of even further fragmentation, the Maryland combat aviation brigade has a Virginia aviation company subordinate to it.

The federal government maintains a semblance of control over the Army National Guard by controlling its funding. Because the Guard is a important cog in the national defense structure, more than 90 percent of its money comes from the federal government.[6] The adjutants general control these funds.[7] To assist the adjutants general, each state has a United States property and fiscal officer, a National Guard officer detailed to extended federal active duty with either the Army or the Air Force. The United States property and fiscal officers ensure that federal funds are obligated and expended according to the appropriate federal statutory and regulatory authorities.[8]

The National Guard Bureau, a joint bureau of the Departments of the Army and Air Force, is charged with controlling federal funding to the state Guards and serving as the channel of communication between the Army and the states and the Air

THE THREE-ARMY SYSTEM

Force and the states on National Guard matters.[9] The chief of the National Guard Bureau, a lieutenant (three-star) general appointed from the National Guard by the president with the advice and consent of the Senate, also serves as the principal staff adviser on National Guard matters to the secretaries and chiefs of staff of the Army and the Air Force. The National Guard Bureau is thus both a staff agency and an operating agency; it does not, however, command or otherwise directly control National Guard units. Within the National Guard Bureau, the director of the Army National Guard, a major general appointed from the Army National Guard, administers Army National Guard programs.

Although the similarities between the Army National Guard and the Army Reserve are to be expected, the two reserve components are strikingly different. In contrast to the Guard, the Army Reserve is an entirely federal force, responsible solely to the Department of the Army. The commander in chief of Forces Command commands all Army Reserve units in the continental United States (less special operations units, which are commanded by the U.S. Army Special Operations Command); Reserve units outside the continental United States are commanded by U.S. Army Pacific or U.S. Army Europe. Unlike the Army National Guard, which consists mainly of combat units, the Army Reserve primarily comprises combat support and combat service support units (table 1 contains the definitions of combat, combat support, and combat service support used throughout this book).

Until just recently, the chief, Army Reserve, a major general appointed from the Army Reserve by the president with the advice and consent of the Senate, served only as a staff adviser to the chief of staff of the Army on Army Reserve matters and commanded no Reserve forces. This arrangement created a vacuum with regard to the Army Reserve, as explained by a senior Army Reserve officer: "Typically, the senior . . . Army leaders and the . . . Army Staff's action officers have tended to pass actions that concern the Army Reserve to OCAR [office of the chief, Army Reserve], even when OCAR had absolutely no functional control over the problems and even though OCAR was in no position to drive either the functionally relevant Army Staff elements or [Forces Command] to arrive at meaningful solutions."[10]

Table 1. Classification of Army Units

Combat Units (directly involved in the conduct of actual fighting)

    Infantry
    Armor
    Field Artillery
    Air Defense Artillery
    Aviation
    Special Operations (some units—special forces, rangers, special operations aviation]
    Engineers (some units—e.g., divisional combat engineers)

Combat Support Units (provide operational assistance to the combat arms)

    Chemical (some units—e.g., division chemical reconnaissance units)
    Engineers (some units—e.g., corps combat engineers)
    Military Intelligence
    Military Police (some units—e.g., divisional military police companies)
    Signal [communications] (some units—e.g., divisional signal battalions)
    Special Operations (some units—e.g., psychological operations and civil affairs units)

Combat Service Support Units (perform personnel service support, logistics, and administrative functions supporting the operations of combat and combat support units)

    Adjutant General [personnel service]
    Supply and Service
    Transportation
    Maintenance
    Signal (some units—e.g., theater communications units)
    Military Police (some units—e.g., enemy prisoner of war units)
    Medical
    Chemical (some units—e.g., corps decontamination units)
    Engineers (some units—e.g., construction units)
    Finance
    Judge Advocate General [legal service]

Source: Robert L. Goldich, *U.S. Army Combat-to-Support Ratios: A Framework for Analysis* (Washington, D.C. Congressional Research Service, 1989).

In 1990, however, after much foot dragging by the Army in response to previous congressional prodding, Congress legislated the establishment of the U.S. Army Reserve Command (USARC) on a two-year test basis.[11] Congress directed the Army to place USARC under the command of the chief, Army Reserve and to assign it as a major subordinate command of Forces Command. Under this law, most Army Reserve units in the continental United States are to be assigned to and commanded by USARC. The Army activated USARC as a provisional command in October 1990, and the command became fully operational in 1992.

Like the Army National Guard, the Army Reserve is organized along geographic lines. The major U.S. Army Reserve commands compose the level of command beneath USARC. There are two types of major U.S. Army Reserve commands. General officer commands, although they consist of subordinate units from the same general vicinities, are functional organizations—that is, they are organized and equipped to perform a wartime mission—that are commanded by general officers. Army Reserve commands (ARCOMs) are Reserve administrative headquarters that exist only to command and control Army Reserve units within their geographic areas. ARCOMs are commanded by Army Reserve major generals. Subordinate to the major U.S. Army Reserve commands are smaller Army Reserve units, although some general officer commands have no assigned subordinate forces, but instead are major tactical headquarters that will be assigned subordinate units based on the wartime missions they receive.

The active component maintains an elaborate structure to support the reserve components (see fig. 3). Directly subordinate to Forces Command are four continental U.S. armies (CONUSAS) commanded by Regular Army lieutenant generals (one CONUSA, Fourth Army, has been recently inactivated). The CONUSAS are not warfighting theater armies (see table 2); aside from their territorial defense mission, they exist solely as administrative headquarters to support reserve units during peacetime and mobilization. The CONUSAS are geographically oriented. Second Army, for example, is headquartered at Fort Gillem, Georgia, and is responsible for reserve units throughout the southeastern United States.

Each CONUSA has several subordinate active component read-

The Future of the Citizen-Soldier Force

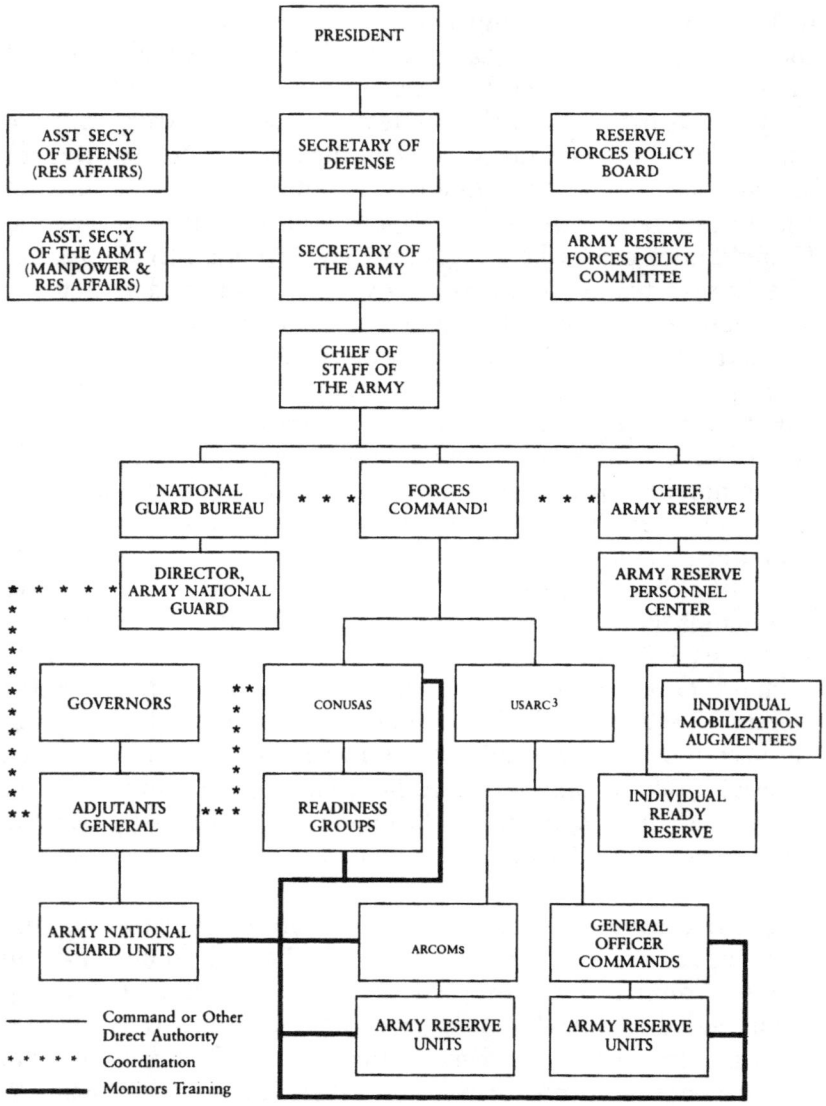

1. Army reserve units outside continental United States are commanded by U.S. Army Europe and U.S. Army Pacific; Reserve special operations units are commanded by U.S. Army Special Operations Command.
2. Also Commander, USARC.
3. CONUSAs retain command authority over ARCOMs and general officer commands until transition complete.

Figure 3. Lines of Authority for the Army's Reserve Components

Table 2. The CONUSAs and Theater Armies

The Continental U.S. Armies (CONUSAS)

Army	Headquarters
1st	Fort Meade, Md.
2nd	Fort Gillem, Ga.
5th	Fort Sam Houston, Tex.
6th	The Presidio, San Francisco, Cal.

The Theater Armies

Army	Location	Function	Commander
3d	Fort McPherson, Ga.	Army Component Headquarters of U.S. Central Command	Lieutenant General
7th	Heidelberg, Germany	Army Component Headquarters of U.S. European Command	General
8th	Seoul, South Korea	Army Component Headquarters of U.S. Forces Korea (a unified command subordinate to U.S. Pacific Command)	General

iness groups, whose mission is to advise and assist reserve component units in achieving the readiness the Total Force Policy demands of them. Commanded by active Army colonels, the readiness groups are small organizations responsible for multistate areas, and habitually have received less than top priority from the Army in terms of both the quality and quantity of soldiers assigned to them.[12] Further, because assignment as a readiness group adviser is tough duty "out of the Army mainstream" (i.e., away from active component troop units), it is not highly coveted.

Although they are in the process of ceding this responsibility to USARC, currently the CONUSAs actually command all Army Reserve units in their geographic areas. Because the Army National Guard is commanded by the states and not directly controlled by the Army in peacetime, the CONUSAs legally are limited to monitoring and evaluating the training of Guard units in their regions. With the establishment of USARC, the CONUSAs apparently will perform the same functions for the Army Reserve that they have traditionally performed for the Army National Guard.

The system described above is undeniably complex. In actuality, it is three separate systems. Separate command structures exist for the active Army, the Army National Guard, and the Army Reserve; the Army commands its active forces directly through its major commands but controls its co-equal reserve forces through USARC and the states. Nonetheless, the Army depends on all three of its components to coalesce and fight as one. The Army system designed to accomplish this operational merger is the CAPSTONE program.

The goal of CAPSTONE is to produce units that "know their wartime chain of command, where they will fight and with whom."[13] The program establishes three separate relationships between active and reserve component units.[14] The first of these is the wartime command relationship. CAPSTONE, based on operation plans, establishes the wartime alignment between reserve units and their wartime gaining commands; that is, CAPSTONE identifies a reserve unit's planned wartime chain of command. A reserve component unit's wartime CAPSTONE alignment may include both superior and subordinate units of all three components—the Army Reserve, the Army National Guard, and the active component (see fig. 4). Thus, a reserve unit may have subordinate active units in its CAPSTONE alignment, and its wartime gaining command may itself be a reserve unit. Some reserve component units, because they are simultaneously programmed into operation plans in different theaters of operations, have multiple CAPSTONE alignments (e.g., one for Europe and one for Korea).

The second relationship formed by the CAPSTONE program is the directed training association. There are several types of directed training associations, but the two most important are roundout and affiliation. Roundout units bring "understructured [active component] units to designated organizational structure."[15] In

THE THREE-ARMY SYSTEM

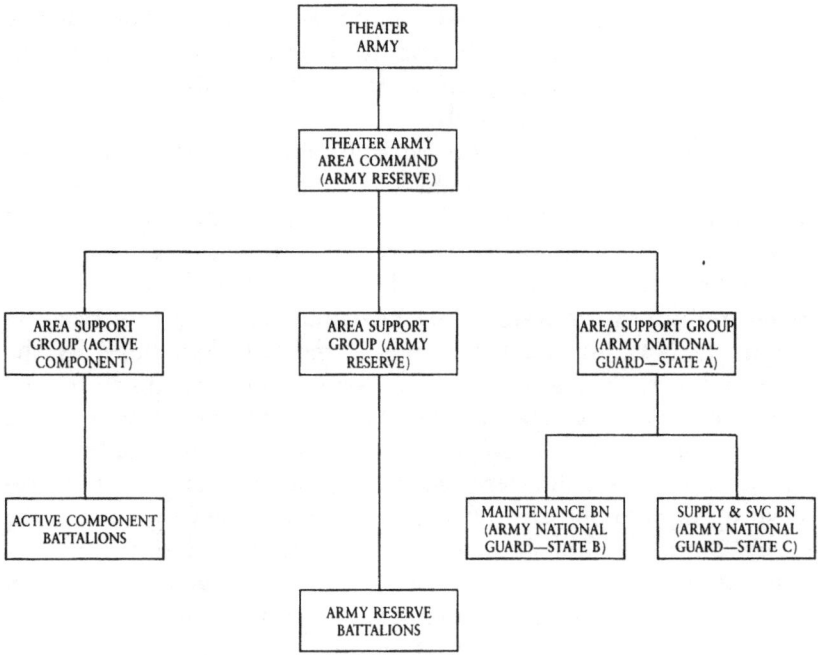

Figure 4. Hypothetical CAPSTONE Alignment

the most common roundout scenario, an active division with three assigned brigade headquarters has two active component brigades and one reserve component brigade. Except for the 6th Infantry Division (Light), whose roundout brigade is an Army Reserve unit, roundout brigades are Army National Guard units. Roundout creates complex command relationships, because as Guard units, the roundout brigades are commanded not by their parent divisions but by the states. As if this state-federal relationship were not complicated enough, many of the Army National Guard roundout brigades have subordinate units from more than one state.

The affiliation relationship is designed to compensate for geographical separation between CAPSTONE-aligned units. The purpose of this relationship is to increase the readiness of reserve component units by pairing in peacetime similar active and reserve organizations that do not necessarily have wartime command relationships. The active component units provide training assistance and equipment support to their affiliated reserve component units; commonly, the active units evaluate the reserve units' annual training periods.

The third, and perhaps most significant, CAPSTONE relationship is the peacetime command and control relationship. This is the Army's attempt to translate the formal wartime CAPSTONE relationship (which, during wartime, it has the power to enforce because reserve component units become part of the active Army when mobilized for war) into a peacetime relationship. Because the Army National Guard in peacetime is commanded by the state and territorial governors, the Army cannot establish formal command relationships between Guard units and either active component or Army Reserve units, or, for that matter, between Guard units from different states. Similarly, because Army Reserve units actually are (or will be) commanded by USARC, the Army cannot establish peacetime command relationships between active component and Reserve units, or even between Reserve units commanded by separate ARCOMs, without revamping the current structure. Even funds earmarked for CAPSTONE (those for reserve component units' participation in large-scale training exercises with the active component, for example) are controlled by the peacetime chain of command; the wartime CAPSTONE chain of command does not even manage its own peacetime money.[16]

In short, with respect to the Army National Guard, the Army is powerless to establish and enforce formal peacetime relationships, and with respect to the Army Reserve, it has chosen not to do so. The peacetime CAPSTONE relationship is thus in reality no formal relationship at all; it is merely a codified informal system of voluntary information exchange. As the Army regulation governing the CAPSTONE program states:

> [CAPSTONE] associations cross existing peacetime command lines. These associations will not circumvent or violate existing command channels on matters of policy, command and control, training management, planning, or funding. For the unit associations to function effectively, extensive coordination will follow established peacetime command channels. However, when appropriate and with the consent of commanders [for the Army Reserve] or adjutants general [for the Army National Guard] involved, informal

coordination may follow the CAPSTONE association channel. . . . Frequent exchanges of information between peacetime and wartime command headquarters are encouraged to insure that subordinate units receive prompt wartime mission planning guidance.[17]

In translation, the CAPSTONE peacetime command and control relationship is an attempt to encourage voluntary coordination during peacetime between units that are slated to fight together during wartime. Under the current three-army system, voluntary cooperation among a multitude of players is the only way the peacetime CAPSTONE relationship can be effectuated.

Conceptually at least, CAPSTONE is the Army program that bridges the gap between the peacetime structure and war plans. Mobilization, controlled by statute, is the mechanism by which the reserves become part of the wartime Army when plans are executed.

There are five categories of mobilization (see table 3). *Selective mobilization* involves the call-up of reservists for domestic purposes.[18] The *200K call-up* empowers the secretary of defense, with presidential authorization, to order units and individual members of the Selected Reserve (which does not include the Individual Ready Reserve) to active duty for an initial term of ninety days "when the President determines that it is necessary to augment the active forces for any operational mission."[19] The secretary of defense, again with authorization from the president, may extend this period an additional ninety days. The total number of reservists on active duty under this statute may not exceed two hundred thousand (hence the *200K*).

Partial mobilization can occur only when the president declares a national emergency.[20] Partial mobilization allows the Department of Defense to mobilize up to one million members of the Ready Reserve for up to twenty-four months, and thus allows mobilization of the Individual Ready Reserve. The final two mobilization categories, *full mobilization* (mobilization of all reserves) and *total mobilization* (mobilization of all reserves plus expansion of the Army through conscription), can occur only when Congress (not the president) declares war or a national emergency.[21] When full or total mobilization occurs, all reservists can be mobilized for the duration of the war or national emergency.

Table 3. Mobilization Categories

Mobilization type	Statutory authority	Action required to invoke mobilization authority	Category of reservists that can be mobilized	Number of reservists that can be mobilized	Duration of mobilization
Selected	10 U.S.C. §§ 331, 332, 3500	Presidential order	All	All necessary	Duration of emergency
200K	10 U.S.C. § 673b	Presidential order	Selected Reserve	200,000	90 days, with provisions for a 90-day extension
Partial	10 U.S.C. § 673	Presidential declaration of national emergency	Ready Reserve	1,000,000	24 months
Full	10 U.S.C. § 672	Congressional declaration of war or national emergency	All	All	Duration of war plus 6 months
Total	10 U.S.C. § 672	Congressional declaration of war or national emergency	All	All	Duration of war plus 6 months

When a reserve component unit is alerted for mobilization, it assembles at its Reserve center or National Guard armory. From there, the unit moves to a mobilization station, normally an active Army post. Generally, the reserve unit conducts its post-mobilization training at its mobilization station and from there is transported to its final destination.

THE THREE-ARMY SYSTEM 21

Selected Reserve units train regularly. Typically, reserve units assemble for thirty-nine days annually. Those thirty-nine days are usually divided into one fifteen-day annual training period (although for specific exercises, such as overseas deployment training, annual training periods are longer) and twelve two-day weekend "drills." A typical weekend drill consists of four *unit training assemblies,* each of which is four hours long. Depending on their training schedules, units may eliminate one two-day drill from the calendar and redistribute the four unit training assemblies to create drill periods of five unit training assemblies that begin on Friday evenings. Special units, such as airborne and aviation units, are authorized additional unit training assemblies.

Although they are fully assembled only during scheduled training periods, reserve component units, like all other units of the U.S. Army, must comply with the various Department of the Army (and National Guard Bureau, in the case of the Army National Guard) regulations, directives, and procedures. Army Reserve units must comply with these mandates because they are subordinate to the Department of the Army; Army National Guard units must comply as a condition of continued receipt of federal funds. These authorities necessitate the accomplishment of certain day-to-day administrative functions, thus requiring some sort of full-time presence in reserve units.

The soldiers who administer reserve component units on a daily basis are *full-time support personnel.* The duties of full-time support personnel relate to all areas of reserve unit operations, including administration, logistics, personnel management, recruiting, reenlistment, and training.

There are three categories of full-time support personnel. Active Guard/Reserve (AGR) soldiers are reserve component soldiers on active duty for at least 180 days. These soldiers are paid from reserve component appropriations. Army Reserve AGR personnel are centrally managed by the Full-Time Support Management Center, a field operating agency of the office of the chief, Army Reserve. Army National Guard AGR personnel, on the other hand, are in practice managed by the states, even though they are subject to the supervision of the Army National Guard Personnel Center and even though they are appointed under Title 10 of the United States Code (the provisions governing the federal armed forces) rather than Title 32 (the provisions governing the National Guard). The state/federal identity of Army National Guard AGR soldiers is therefore ill defined, although practically speaking these soldiers are responsible to the states, just as their part-time Guard counterparts are. AGR personnel receive the same pay and benefits as soldiers of like grade in the active component, and AGR soldiers must meet the same standards as active-duty soldiers, making AGR and active component soldiers, on the surface, at least, virtually indistinguishable.

Military technicians are the second class of full-time support personnel. Technicians are federal civil service employees who are required, as a condition of their employment, to be members of their Selected Reserve units. Thus military technicians, even though they wear Army uniforms to work each day, are civilians except during unit training periods; the difference between technicians and the part-time soldiers they support is that the technicians' civilian jobs are at the Army Reserve center or National Guard armory. Because civil service pay scales tend to be lower than military pay scales, and because civil service employees are not entitled to military benefits, technicians cost less per capita than AGR personnel.

The final category of full-time support personnel consists of active component soldiers who directly support reserve component units—not readiness group advisers, but soldiers who are actually assigned to reserve units or who work with and in those units day to day.

An overview of the reserve components in the context of today's political climate would be incomplete without at least a cursory description of the fiscal and budgetary implications of the reserve components vis-à-vis the active component. Until 1982, the Army National Guard and Army Reserve depended on the Department of the Army to allocate reserve component procurement funds from the Army budget appropriated by Congress. This system of resource allocation impeded the reserves' readiness through, as one study concluded, "a neglect of the reserves in Pentagon planning, programming, and budgeting, processes that [were] controlled largely by active-duty military officers."[22] The reserve components, for example, were equipped in large part with leftover active component equipment. In 1982, Congress enacted legislation mandating that the Army National Guard and the Army Reserve be equipped through the expenditure of funds earmarked specifically for the reserves. Each of the three Army components now maintains separate procurement accounts, as well as separate personnel, operations and maintenance, and military construction accounts. Some 14.3 percent of the Army budget goes to the reserve components—9.8 percent to the Army National Guard and 4.5 percent to the Army Reserve.[23]

The House Armed Services Committee has estimated that a reserve unit costs from 40 to 70 percent of an active component unit.[24] There are three major reasons for this cost differential: lower direct personnel costs, lower personnel support costs, and lower operating tempo (training time and equipment usage).[25] Obviously, because the majority of paid reserve personnel are on duty only one weekend per month and an additional two weeks annually, the reserve component payroll is less per capita than the active component payroll, even though a reserve soldier is paid a full day's base pay for one four-hour unit training assembly or four days' pay for a two-day weekend drill. Thus the typical drilling reservist is paid the equivalent of sixty-two days' pay annually, approximately one-sixth of the annual pay of an active-duty soldier.[26]

Personnel support costs are lower because part-time reservists, with the exception of their annual training periods, are not entitled to many of the benefits active Army soldiers receive, such as free military medical care. Additionally, the active component is an essential source of personnel for the reserves. Because soldiers who move from the active Army to the reserve components have been previously trained, reserve component individual training costs are lower than those of the active component.

Finally, because the typical reserve unit generally trains only thirty-nine days annually, the reserve components use their equipment less. Taxpayers spend less per year on fuel, ammunition, and spare parts for a reserve unit than for a like active unit. As an estimate, in terms of annual operating expenditures, a reserve tank or mechanized infantry unit costs 25 percent of a similar active unit.[27]

In sum, the Total Force Policy mandates the equality of the Army's three components, but, even on the surface, the components remain separate. Three disparate peacetime structures, each resourced separately, exist to accommodate these three separate armies. The active component is organized functionally, as in wartime; the reserve components are organized geographically. The channel of communication for the active component in peacetime generally flows directly through the functional chain of command, just as in wartime. For the reserves, however, the peacetime channels of communication are through administrative headquarters: USARC and the ARCOMs for the Army Reserve and the National Guard Bureau and the states for the Army National Guard.

The current systems are the products of tradition, politics, and administrative convenience (some would argue necessity). The remainder of this book examines the origins of those systems, asks some tough questions about them, and proposes some answers to those questions. I make no apologies for finding fault, but finding fault and pointing fingers are not my purposes. As a nation, we deserve an Army—both a standing professional army and a citizen-soldier force—that is capable of defending our democratic institutions given the international and domestic political climates that will exist in the foreseeable future. As an Army, we need a reserve force that measures up to the standards of the citizens we defend—and of the taxpayers who pay for that force. My purpose here is to suggest some paths to those goals.

The Total Force Policy encompasses the reserve components of all the services, but this book is about only the Army's reserve components, the Army National Guard and the United States Army Reserve. Nonetheless, for simplicity's sake, I have continued to use the term "Total Force Policy." Additionally, although the Army National Guard and the Army National Guard of the United States are, strictly speaking, distinct organizations, for purposes of this book, that distinction is not altogether significant, and I refer to both entities as the Army National Guard.

In the same vein, many of my proposals would, or could, affect the other services were they to be implemented. Reforms of the Army National Guard, for example, implicate the Air National Guard as well. I do not analyze those effects. Obviously, such a global analysis is a prerequisite to the implementation of any serious reforms of the Army's reserve components.

As any analysis must, this book deals in generalities. Of

course, there are always exceptions. I have not pointed out every one. Policies must be based on the rules, not on the exceptions.

Finally, my intent most certainly is not to impugn the integrity, dedication, or professionalism of any Army National Guard or Army Reserve soldier. Not enough can be said about the men and women who work full-time civilian jobs to pay the rent and trot off to war upon receiving a phone call. These soldiers are, however, products of a system (or several systems, to be more accurate), and the deficiencies in that system need to be pointed out. There has been enough chest thumping about how good our reserves are and have been. The focus of this book is on problems; self-aggrandizement in peace does not win wars.

The issues this book discusses have always been fraught with emotion, but emotion has never led to sound national defense policies—indeed, it has led to the retention of many *unsound* military policies. No doubt many will disagree with the book's conclusions; certainly, those conclusions are controversial and politically volatile. Realizing the stakes, however, the disagreement hopefully will foster spirited debate and not blind defensive volleys, and my contribution will have been to provide food for thought rather than targets for the political rifle range. After all, when we shoot on that range we shoot only at ourselves, and there are no winners.

2. THE EVOLUTION OF THREE ARMIES

> History is relevant, and officials responsible for shaping the nation's mobilization policies should have studied it. Instead, avoidable mistakes were repeated by several generations of decisionmakers and planners, who share some of the responsibilities for the poor showing of the Army reserves since the Second World War.
> —Martin Binkin and William W. Kaufmann

HOW has our nation, with an army that is the most technologically advanced in the world and that has developed the world's most innovative warfighting doctrine, managed to end up with such a cumbersome reserve system? The answer is simple: history and politics. Tradition and politics are the bedrock of the reserve components. Without an understanding of these factors, we cannot see beyond the tip of the iceberg.

A historical sketch of the Army National Guard and the Army Reserve illuminates several interesting points for today's analysts, planners, and would-be reformers. Many recent "innovations" and "revolutionary ideas" are not really new. The problems that bedevil the Army with respect to the citizen-soldier force today—the tension between the states and the federal government, the limitations imposed by geography, the inability of the reserves to meet expectations fully when mobilized, the air of distrust between the citizen-soldier and the regular, and the pervasive political influence of the reserves, to name but a few—have existed from the inception of the militia system.

Without understanding the history of the reserves, one cannot begin to comprehend the factors that weigh in the balance today with regard to either the current capabilities of the reserve components or prospects for improvement. This chapter highlights the historical background of the specific problems addressed in later chapters. By no means is the chapter intended as a detailed or original historical analysis. This book analyzes the present, not the past, and, in any event, analyses of the 350-year history of the reserves are the subjects of books, not chapters. Accordingly, this brief historical sketch is necessarily and intentionally selective, but it is nonetheless essential.

The Army National Guard, the senior of the Army's two reserve components, traces its beginnings to the English militia tradition, which the colonists brought with them to the New World. The thirteen colonies saw the need to form some type of self-defense force, and accordingly, the colonies imposed military obligations on their able-bodied male citizens. The colonial militia centered on companies formed in townships, counties, or cities.[1] Thus began the tradition of geographically oriented citizen-soldier units, a tradition that continues to pervade our reserve system today.

Each colony maintained its own militia. In fact, the pre-revolutionary militia was not one militia but thirteen separate militias, each governed under a different militia system and each reflecting a different culture. True to the English tradition, the militia was limited to fighting only within its own colony, for militia units were to be used for defensive purposes only. This limitation inevitably created problems when the colonies attempted to wage warfare collectively. In 1754, for instance, George Washington was foiled in his attempt to employ the Virginia militia in the Ohio country because the militiamen would not cross the colonial border.[2]

In general, the colonial militias were parochial organizations that did not cooperate with one another. Combining units from these disparate forces to form larger units was difficult under these circumstances; nevertheless, the militia was the mainstay of the colonial defense force.

After the Revolution, the militia was governed by the Articles of Confederation. Article 9 provided: "Congress . . . shall have authority . . . to agree upon the number of land forces, and to make requisitions from each State for its quota, in proportion to the number of white inhabitants in such State; which requisitions shall be binding; and thereupon, the legislature of each State shall appoint the regimental officers, raise the men and cloathe, arm, and equip them in a soldier-like manner, at the expence of the United States." Under the Articles of Confederation, therefore, the federal government could raise troops only by requesting the states to provide them. This system proved inadequate because the Congress could not force the states to honor its requests, and the states were often uncooperative.[3]

At the constitutional convention, the framers of the Constitution sought to improve this system. Competing considerations, however, drove their final solution. The framers accepted the necessity of a national defense system, but, heedful of the abuses they had suffered at the hands of the British army, many of them sought to avoid creating a professional standing army that potentially could become too powerful to control. Accordingly, they desired to create a regular force that was no larger than absolutely necessary and to place a significant portion of the national defense burden on the militia.

The constitutional provisions governing the militia were a compromise between the framers who wanted the states to exercise full dominion over the militia and those who desired greater federal control. The "militia clauses" of the Constitution provide:

> The Congress shall have Power . . .
>   To provide for calling forth the Militia to execute the laws of the Union, suppress Insurrections and repel Invasions;
>   To provide for organizing, arming, and disciplining, the Militia, and for governing such Part of them as may be employed in the Service of the United States, reserving to the States respectively, the Appointment of the Officers, and the Authority of training the Militia to the discipline prescribed by Congress.

In 1792, Congress passed the Militia Act, which implemented the militia clauses of the Constitution (and which remained in effect until the twentieth century). The act mandated that every free, able-bodied white man would serve in the militia and would arm and equip himself. Because the act maintained the existing separate state militias, Congress consciously washed its hands of the militia, abdicating the governance of the citizen-soldier forces to the states.

The postrevolutionary militias were dominated by diversity and lack of organization. The influence of geography remained prevalent: on the frontier, where the militia was most needed, units were hard to form because of the sparsity of the population; units in the more densely populated cities tended to be more cohesive. The militia served the states primarily, the United States only secondarily.

Even in the early nineteenth century, politics began to inject itself into the American militia system. Some militia units began to take on partisan political characteristics; there were Federalist units and Democratic units. When John Adams became president, Republican militia units refused to escort him through Philadelphia.[4]

The inefficiency of the militia clauses and the act implementing them manifested itself regularly through the ineffectiveness of the militia as a fighting force. In 1812, faced with the imminent threat of British invasion, Congress authorized the alert of one hundred thousand militiamen. The governor of Connecticut, however, called this action unconstitutional, opining that the cruising British fleet did not constitute an invasion or insurrection. He refused to provide his state's militia to the federal government. The governor of Massachusetts similarly refused. In his *History of the Militia and the National Guard*, Mahon notes that "the reaction of these two states to the president's call underscored a problem created by the Constitution. Each militia had two commanders-in-chief, a governor and the president, and if those two disagreed on policy, the several militias could not function as a national reserve."[5]

Gubernatorial interference notwithstanding, many miltitiamen simply did not respond to the federal calls to arms. Their loyalties clearly lay with their states. Nor did the militiamen who responded to the nation's call "cover themselves with glory" in the War of 1812.[6] As Major General Henry Dearborn prepared to cross into Canada at the Niagara River, the New York militia under his command refused to cross the border. Incursions beyond the borders of the United States, these militiamen believed, violated the militia clauses of the Constitution. After all, the constitutional mandate to repel invasions "provided no clear sanction for militia to become invaders themselves."[7] Incidents such as this occurred several times throughout the war, demonstrating that the "militia could not be used as an offensive force."[8]

The War of 1812 also saw the beginning of the tension that has existed since then between the citizen-soldier and the regular. The regulars disdained the militia, sometimes with good cause.

Regular officers frequently refused to subordinate their units to militia officers of senior rank. For their part, the militiamen lamented their lack of recognition, seeing themselves as mere cannon fodder to glorify the regulars.

Several states protested the subordination of their militia units to regular officers. A convention in Connecticut declared that such action was unconstitutional. In an attempt to rectify the situation, Secretary of War James Monroe proposed a one-hundred-thousand-man national reserve independent of the states; Congress, however, refused to enact his proposal.

Although the performance of the militia during the War of 1812 was generally poor, several regular officers, among them Winfield Scott and Zebulon Pike, were able to coax the militia to rise to the challenge. The War of 1812 demonstrated the general ineffectiveness of the militia as a national fighting force, but, importantly, it also demonstrated that regulars who understood the capabilities and limitations of the militia could employ citizen-soldiers effectively.

John C. Calhoun, secretary of war from 1817 to 1825, was one of the most influential statesmen of the early nineteenth century with regard to the nation's military affairs. Calhoun argued that the Regular Army, and not the militia, should be the bulwark of the nation's defense force; the militia, he asserted, was ill prepared to fight the professional armies of Europe. Calhoun proposed an expansible army, a cadre of Regular Army officers that would be supplemented by recruits when war broke out. His views of the defense establishment affected the United States Army profoundly for the next one hundred years.

During the campaigns against the Indians, the themes that had manifested themselves during the War of 1812 remained evident. Regular officers continued to look with scorn on the citizen-soldiers, considering them unprofessional, and, indeed, the militiamen often proved that they lacked basic soldier skills. The militia continued to demonstrate its aversion to taking orders from regular officers. During this period, "eventually even Congress awoke to the necessities of the situation, and, grudgingly, increased the Regular Army." [9]

The United States fought the Mexican War largely with federal volunteers; the militia, because of the perceived constitutional limitations, was "not suitable." [10] These volunteers, however, bore the mantle of the citizen-soldier tradition. Politics were conspicuous among the volunteer officers and engendered ill will between the citizen-soldiers and the regulars. Many volunteer officers received their commissions through political connections; President Polk's law partner became a general at thirty-five, an age at which a regular officer likely would still be a lieutenant. President Polk even planned to appoint a political ally as a lieutenant general so that he would outrank Major Generals Zachary Taylor and Winfield Scott, both regulars and both members of the opposing political party.[11]

When the Civil War broke out, President Lincoln immediately called seventy-five thousand state militiamen into federal service. For the remainder of the war, though, the secretary of war assigned quotas to the states and the governors administered a system of volunteers.[12] Citizen-soldiers fought on both sides during the Civil War.

In both the Union and Confederate armies, the historical tension between the regulars and citizen-soldiers resurfaced.[13] In the Union army, "commissions as general officers were given to a host of civilians now principally famous for their misadventures."[14] Tension between the states and their respective central governments also arose, especially in the Confederacy. Both Arkansas and Georgia refused to place their troops under the command of Confederate officers.

The Union militia generally was of poor quality. One astute commentator summed up the performance of the militia during the Civil War by writing, "The common militia was an impractical political concept, the [federal] Volunteer an inescapable reality."[15] Nonetheless, the citizen-soldiers who fought during the Civil War had a profound effect on the American populace after the war: "They never forgot their war experience and never let the country forget it."[16]

By the 1870s, most state militias had adopted the name National Guard, first used by the 7th New York Regiment, but "in spite of the word National, the Guard units were primarily instrumentalities of the states."[17] In 1877 the militia's new name gained national credence when, at a national meeting of organized militia officers, the National Guard Association was formed. The National Guard Association insisted that the National Guard was an integral component of the nation's defense force, and it fought "the development of a monolithic military establishment under the control of regular officers."[18] The National Guard Association immediately undertook to further the interests of the militia through lobbying and other political activities, and always found support for its views from states' rights advocates in Congress.

At about the same time as the formation of the National Guard Association, Emory Upton emerged as one of the era's most influential military critics. Upton was a decorated West Point graduate who had risen during the Civil War to the rank of brevet major general by age twenty-five. He set forth his ideas for reforming the United States Army in *The Military Policy of the United States,* which was not published until 1904, some twenty years after his death, and which was considered for decades the standard historical work on the United States Army.[19]

Upton argued that the "militia system" ought to be abandoned in favor of a regular army patterned after that of Germany. He indicted the militia system, asserting that the first cause of the nation's military weakness was "the employment of militia and undisciplined troops commanded by generals and officers utterly ignorant of the military art."[20] Upton urged the adoption of Calhoun's concept of an expansible army, a force built around a nucleus of regulars and filled by citizens who were products of universal military training. Upton argued that the central government should control all military forces; state control of soldiers intended to function as a national reserve "compounded their uselessness."[21]

But Upton's arguments ran even deeper. He asserted that the major institutional flaw of the American military was civilian control.[22] Upton believed that the German military model, in which the military was controlled by a general staff of professional soldiers and remained free of civilian legislative meddling, was the answer to preventing the recurrence of past American military failures, as he perceived most of the country's previous military operations.

Upton's ideas were widely accepted throughout the Regular Army of that era. However, in the words of Russell Weigley, preeminent historian of the United States Army, "Emory Upton did lasting harm in setting the main current of American military thought not to the task of shaping military institutions that would serve both military and national purposes, but to the futile task of demanding that the national institutions be adjusted to purely military expediency. By proposing a military policy that the country could not accept, Emory Upton helped ensure that the country would continue to limp along with virtually no military policy at all."[23]

War with Spain broke out in 1898, and in many ways this conflict underscores Professor Weigley's insight. The Spanish-American War was also a watershed in the history of the National Guard. As in virtually every previous conflict, citizen-soldiers formed the bulk of the American fighting force. To circumvent the perceived constitutional restrictions on foreign service that had plagued the militia throughout the nineteenth century, National Guardsmen were sworn into federal service as individual volunteers, rather than as state militia units. Integrity of the state Guard units was retained to some extent, highlighting the disparity among the units of different states.

The traditional animosity between the regulars and the citizen-soldiers was again readily apparent. The regulars looked down on Guard officers "as home-town boys, whom no amount of military courtesy could transform into anything more lofty."[24] Indeed, as in the Mexican War and the Civil War, many of the Guard's generals were appointed based on their political connections rather than on their civilian qualifications. Some of the regulars' criticism of the Guard was justified; the 71st New York Regiment broke under fire at San Juan Hill and refused to move forward.[25] The Guardsmen, on the other hand, were convinced that they were equal to the regulars on the battlefield, despite their apparent lack of discipline. In some respects, they too were correct, as evidenced by the performance of citizen-soldier units in the Philippines.[26]

After the Spanish-American War, Elihu Root became the secretary of war. Root, heavily influenced by the ideas of Upton, set out to reform the militia system. Realizing the sheer folly of the nation's reliance on the Militia Act of 1792, which was still in effect, Root stated:

> It is really absurd that a nation which maintains but a small Regular Army and depends upon unprofessional citizen soldiery for its defense should run along as we have done for one hundred and ten

years under a militia law which never worked satisfactorily in the beginning, and which was perfectly obsolete before any man now fit for military duty was born. The result is that we have practically no militia system, notwithstanding the fact that the Constitution makes it the duty of the federal Congress "to provide for organizing, arming, and disciplining the militia," and "for calling forth the militia to execute the laws of the union, suppress insurrections, and repel invasions."[27]

Also concerned with the Army's overall poor performance in the Spanish-American War, Congress held hearings to consider ways to prevent a recurrence of that performance. Congressman Charles Dick, chairman of the House Committee on Military Affairs and a major general in the Ohio National Guard, appointed a committee to draft a bill to present to the annual meeting of the National Guard Association.

The War Department also drafted a bill, which was similar to the National Guard Association bill but which contained a provision that would have created a one-hundred-thousand-man national reserve independent of the Guard. The National Guard Association and the Guard, of course, opposed this provision; they saw it as an attempt to dethrone the Guard as the nation's reserve force. Although the House passed the War Department bill, Root, reading the political tea leaves, withdrew the national reserve provision to save the remainder of the bill while the Senate was considering it. The bill was enacted into law in 1903 and became known as the Dick Act.

The Dick Act increased federal influence over the Guard. It provided for an "Organized Militia," which would be known as the National Guard, and mandated that the Guard would be equipped by the federal government. The act also provided that the Guard would conform to Regular Army organization, would be trained by Regular Army instructors, and would be required to conduct a specified number of armory drills. By including a provision making the Guard the nation's primary military reserve, the act significantly enhanced the Guard's stature.

Notwithstanding these significant reforms, the Dick Act did not alter the distinctly state character and complete state command of the Guard; "the sponsors of the legislation," notes Frederick Wiener, "were disinclined to depart from the traditional militia limitations."[28] Indeed, the House stated that the militia "was never designed to be a militia of the United States, nor under the control of the President, except when called into actual service under some one of the [constitutionally] enumerated contingencies."[29]

The Dick Act did not provide any mechanism for federalizing Guard soldiers for contingencies other than those specified by the militia clauses. Thus Elihu Root considered the National Guard's primary task to be training men in peacetime who would volunteer for federal service during war.

The transition of the National Guard from a social and political club to a military force was not a smooth one. Absence from training was common. Moreover, the War Department itself caused turbulence within the Guard by excessively reorganizing it; from 1903 to 1913, the department disbanded 792 Guard units and formed 902 new ones. The Dick Act did nothing to curb the internal politics of the Guard, and Guard units continued to elect their own officers.

Regular Army officers continued Upton's quest to establish a national reserve free of state ties. These officers believed that national security was unsafe in the hands of military units that were responsible to two commanders in chief (the governors and the president) and two chains of command: notwithstanding the improvements brought about by the Dick Act, the governors could still veto national policy through their control of the militia. Moreover, it remained an unanswered question whether Guardsmen, governed by the militia clauses, could serve outside the United States.

In 1908 Congress passed legislation that enabled the president to federalize Guard units for service outside the United States. In 1912, however, the attorney general—supported by the Army, which preferred the creation of a national reserve over an increase in importance of the National Guard—declared that the law was unconstitutional. Congress also enacted legislation in 1908 that created a Medical Reserve Corps of a few hundred physicians, which ultimately evolved into the United States Army Reserve.

One of the most influential proponents of a national reserve during the early twentieth century was Major General Leonard Wood, the Army chief of staff, who called the Guard "an uncoordinated army of fifty allies."[30] In 1912, Wood and Secretary of

War Henry L. Stimson convinced Congress to authorize the creation of a national reserve, a pool of individuals with prior military service. The resulting legislation required men to enlist in the Army for seven years and to serve the last four in the reserve. Wood's idea failed, however, because the enlistees largely ignored the law; it contained no sanctions.

In 1915, Lindley Garrison, Stimson's successor as secretary of war, again proposed the creation of a national reserve. Garrison's plan relegated the National Guard to a third-string role behind the Regular Army and the Continental Army, his proposed federal reserve. The War Department argued that the National Guard could never be fully responsive to the needs of the federal government as long as the Guard remained under the purview of the militia clauses.[31] President Woodrow Wilson eventually endorsed Garrison's plan.

The National Guard Association opposed Garrison's plan and ardently cultivated congressional opposition to it. Representative James Hay, the influential chairman of the House Military Affairs Committee, proposed an alternative plan that would strengthen the National Guard by increasing federal influence over it. Under the Hay bill, the federal government would equip, train, and pay the National Guard. Political patronage would be curtailed through a provision allowing the federal government to approve state officers for concurrent federal commissions. Finally, Hay's plan would require Guardsmen to swear an oath to respond with their entire units to federal calls for service anywhere, assuring federal control of state militia units during war.[32]

President Wilson, bowing to political pressure, finally withdrew his support for Garrison's plan and endorsed the Hay bill. The House passed the bill, but the Senate adopted a plan that revived the Continental Army. On the eve of World War I and as the Texas, New Mexico, and Arizona National Guards were called up to fight Pancho Villa along the Mexican border, the House and Senate compromised and passed the National Defense Act of 1916.

The 1916 act, significant portions of which remain in effect today, adopted much of Hay's plan for the Guard. The legislation broadened the scope of federal control over the National Guard, increasing federal funding and giving the secretary of war authority to cut off federal funds to states whose units failed to comply with federal regulations. The act created the Militia Bureau, headed by a Regular Army officer, within the War Department and instituted the current system of forty-eight armory drills per year (i.e., one weekend per month) and a fifteen-day annual training period.

Although the act provided a solution that allowed Guardsmen to serve overseas at the behest of the federal government, it did not solve completely the problem of Guardsmen serving overseas. The act empowered the president to draft *individual* National Guardsmen into federal service, but provided that those individuals would be discharged from the state militia for the duration of their service in the federal armed forces.

By securing its position as the primary federal reserve force, the 1916 act clearly granted the Guard its wish. The act did, however, create a wholly national reserve. The Officers' Reserve Corps (which soon subsumed the Medical Reserve Corps) and the Enlisted Reserve Corps were formed as manpower pools of soldiers with prior service in the Regular Army. The act also created the Reserve Officers' Training Corps (ROTC) to train and commission Reserve officers at colleges and universities.

As the United States prepared to enter the First World War, it had, "without really knowing it, opted for a citizen army."[33] Despite the opposition of the Uptonians—Root, Garrison, and Wood, among others—the National Guard and the National Guard Association had wielded enough political power "to see [the National Guard] established as the principal part of such reserve as the nation had."[34]

When the United States plunged into the World War, the president drafted the National Guard into federal service. Once drafted, Guardsmen became individual members of the United States Army and lost any legal connection with the state National Guard. To replace the militia forces they had lost to the federal government, the states created State or Home Guards; the War Department subsequently legitimized these organizations by constituting the United States Guard.

National Guard units eventually proved themselves up to the task in World War I. Like all other Army units that fought in France, though, the Guard units received extensive training after their federalization, both in the United States and in France, before their commitment to combat. The prolonged postmobilization training the Guard underwent clearly contributed to its battlefield success.

The Organized Reserve also contributed significantly in World War I. By June 30, 1917, the Enlisted Reserve Corps numbered approximately fifty-five thousand men and provided many signal, engineer, and other support units. The Officers' Reserve Corps provided nearly ninety thousand officers during the war.[35]

As in wars past, animosity between the citizen-soldier and the regular reemerged during World War I. Guardsmen maintained that the system the Army used to number its regiments, which did not retain the original and often historic state regimental designations, was purposely designed "to destroy the identity of the Guard units and in other ways to discredit the Guard."[36] General Peyton C. March, the chief of staff of the Army and a devotee of the Upton school, declared that there was only one United States Army, and ordered Guardsmen to remove from their uniforms any insignia that indicated state affiliation. The Army relieved many senior National Guard officers and replaced them with regulars; this practice seemed to some "to be more a device to bring in favored regular army replacements than to improve combat efficiency."[37]

At the end of World War I, proposals to establish a national reserve again surfaced, along with proposals for universal military training. One of the leading advocates for a national reserve was Colonel (later Brigadier General) John McAuley Palmer, a West Pointer and protegé of General of the Armies John J. Pershing. Palmer, although a Regular Army officer, believed strongly in the citizen-soldier concept and is often called "the father of the Army Reserve."

These proposals would have shrunk the role of the National Guard in the military establishment, and, predictably, the National Guard Association opposed them. The National Guard Association lobbied for the formal designation of the Guard as an integral component of the federal military, and for a scheme of mobilization that would bring Guard units, rather than individuals, into federal service during national emergencies. "The [National Guard Association] made its points where they counted, to senators and representatives, many of whom were conditioned to favor a state-connected reserve system. . . . The [National Guard

Association], cooperating closely with the Adjutant Generals' Association (comprising the states' adjutants general), worked tirelessly to keep the regular army small and to build up the Guard."[38]

The Guard lobby saw most of its efforts realized in the National Defense Act of 1920, including its insistence that the Militia Bureau be directed by a National Guardsman, but it failed in its attempt to have Congress legislate unit integrity upon mobilization. To the Guard's dismay, however, in the 1920 act, Congress also adopted most of John McAuley Palmer's ideas (but not all of them; Palmer was a staunch advocate of universal military training, which was politically unpalatable). Significantly, the act defined the Army as consisting of the Regular Army, the National Guard, and the Organized Reserves. The act also maintained the Officers' and Enlisted Reserve Corps, and although these entities continued to exist as manpower pools, provisions were made to form units in the Organized Reserves.

As the Organized Reserves became more prominent, the Reserve Officers Association was formed in 1922 as the Reserves' counterpart to the National Guard Association. In 1923 the Army created the position of executive for Reserve affairs on the Army General Staff. The officer filling this position exercised staff supervision over reserve matters.[39] (In 1941, the position became the executive for Reserve and ROTC affairs; in 1954, the title of the position was changed to chief, Army Reserve and ROTC affairs. Finally, in 1967, Congress legislated the creation of the chief, Army Reserve.)

In 1933, Congress passed legislation that affected the nation's reserves more significantly than any since the Militia Act of 1792. The National Defense Act of 1933 made the Guard "a part of the Army at all times" and ensured that its units would remain intact when federalized.[40]

The National Defense Act of 1933 conferred a dual status on the Guard by creating an entirely new organization—the National Guard of the United States. In its state capacity, the National Guard continued to be governed by the militia clauses of the Constitution. As the National Guard of the United States, however, the Guard was a reserve component of the U.S. Army, to be administered under the clause of the Constitution that empowers Congress to "raise and support armies"—the "army clause." Colonel Frederick B. Wiener, one of the Army's foremost legal scholars, wrote in 1940 that the National Defense Act of 1933 "placed the final mark of inadequacy on the militia clause. For the 1933 Act proved conclusively that a well-regulated militia is impossible of attainment under the militia clause, and can be organized only by resort to the plenary and untrammelled powers under the army clause."[41]

Congress designed this dual system to avert the recurring problems surrounding the use of the militia outside the United States and to obviate the necessity of drafting militiamen individually into federal service. Under the new system, the Guard could be *ordered* into federal service as units under the army clause when Congress declared a national emergency. When ordered into federal service, the Guard would lose its identity as the state militia and don its hat as a reserve component of the Army. Thus "every Guardsman [became] a reservist as well as a militiaman."[42]

Despite the reforms of the 1933 act, the National Guard still found itself caught between the state and federal governments. During the Depression, for example, several governors used National Guard troops to block implementation of New Deal programs.

Other historical problems continued in the period between the world wars. The regulars still looked askance at the Guard. The top Army leadership "saw scant potential in the National Guard" and did not take the steps necessary to ensure its readiness. The Guard was chronically short of equipment. Likewise, many Officers' Reserve Corps units (because the Enlisted Reserve Corps had dwindled, Organized Reserve units generally consisted only of officers) did not even have weapons, and unlike the Guard, which was guaranteed paid drills by the National Defense Act of 1916, funds for the annual training of even a substantial minority of the Officers' Reserve Corps were not appropriated.[43] The Army also continued its disruptive reorganizations of National Guard units; at least seven hundred units underwent some type of organizational change.[44]

With war brewing again overseas, Congress declared a national emergency on August 27, 1940, authorizing the president to order the Guard into federal service under the provisions of the National Defense Act of 1933 and to mobilize the Organized Reserves. By June 1941, the National Guard was fully federalized, and more than forty-six thousand Officers' Reserve Corps officers were on active duty.[45] As it had done in World War I, Congress authorized the states to create new militia organizations to replace the federalized National Guard.

The federalization of the Guard accomplished the chief objective of the National Guard and the National Guard Association: the Guard was finally being used as the nation's primary reserve. The Regular Army, however, was dissatisfied with the Guard's performance. Lieutenant General Lesley McNair, chief of staff of the Army's General Headquarters, wrote that "the National Guard, as organized before the war, contributed nothing to national defense. . . . The structure of the National Guard was pregnant with disaster for the entire nation."[46] McNair charged that Guard generals were uniformly incompetent. Many Guard junior officers were in their forties, and, for this reason as well as others, many were removed.

The Guard, of course, vehemently disputed McNair's charges. It asserted that the Regular Army schemed to remove Guard officers to satisfy "a volcano of erupting military ambitions" among regular officers.[47] One historian has asserted that the Regular Army made the National Guard the scapegoat for the Regular Army's own lack of preparedness.[48]

As was the case in World War I, National Guard units performed creditably on the battlefield in World War II. As was also the case in World War I, however, those units were the beneficiaries of extensive training before being committed to combat. Moreover, the Guard units were diluted with draftees, although most managed to retain their character as regional organizations.

The Organized Reserve also contributed greatly to the Army in World War II. The Officers' Reserve Corps eventually provided 180,000 officers during the war, almost one-quarter of the Army's total officer strength.[49] The prewar Officers' Reserve Corps units, however, were disregarded, and Organized Reserve units, as such, did not serve during the Second World War.[50]

After the Second World War, the National Guard reverted to its peacetime political persona. The National Guard Association persuaded Congress to appropriate one million dollars annually for a public relations campaign to bolster the Guard's image. In

1947, the Guard began placing full-page advertisements in the leading periodicals in the country. It also presented a weekly radio program.

The Regular Army-National Guard bickering continued unabated. National Guard Association officials often testified before Congress, and often berated the Army's top leaders. Guardsmen were convinced that the Regular Army, in the pattern of Calhoun and Upton, sought to oust the National Guard from the place it had earned in the military establishment.

As for the Reserve, the postwar Officers' Reserve Corps numbered nearly 247,000 in 1947. The Enlisted Reserve Corps, on the other hand, numbered less than 470,000 in 1948; only 3.2 percent of discharged enlisted veterans chose to retain their military affiliation.[51]

In 1947, James Forrestal, the secretary of defense, appointed a board to study the nation's reserves. Named for its chairman, Gordon Gray, an assistant secretary of the Army, the Gray Board concluded that the dual status of the National Guard was detrimental to national security and stated that "the National Guard must be directly under Federal control."[52] The Gray Board recommended merging the National Guard with the Organized Reserve Corps, and Secretary Forrestal concurred.

Vociferously opposing the Gray Board's recommendations, the National Guard Association immediately went to work. It counterattacked by charging that the Regular Army was up to its old tricks. The Pentagon, it asserted, through "innuendos and falsehoods," was trying to "divide and destroy the National Guard."[53] The association "hurried with the matter to [its] ultimate protector, the Congress of the United States,"[54] and Congress, harkening to the Guard and its states' rights message, quashed the proposed merger.

During the war in Korea, about one-third of the Army National Guard was mobilized (the *Army* National Guard was born when the Air National Guard was established in 1945). Of these troops, only two Army National Guard divisions were deployed to Korea, and they were deployed more than one year after mobilization.[55] During the Korean conflict, the Army adopted an individual rotation policy; soldiers were rotated home after one year in the combat theater. The deployed National Guard divisions, therefore, eventually lost most of their hometown identities, because, when their initial cadre of Guardsmen went home, they received draftees as replacements.

The Organized Reserve Corps contributed even more manpower to the Korean War effort than did the Guard. In all, 240,500 Organized Reserve Corps members were activated, and fourteen separate Organized Reserve Corps battalions and forty separate Organized Reserve Corps companies deployed to Korea.[56] Although they eventually performed well, the reserve units and personnel mobilized for the Korean conflict initially were largely unprepared.[57]

In 1952, Congress enacted the last substantial reform of the nation's reserve system, the Armed Forces Reserve Act of 1952. The act created the current reserve categories—the Ready Reserve, Standby Reserve, and Retired Reserve. The legislation also officially eliminated the Officers' Reserve Corps and the Enlisted Reserve Corps, and renamed the Organized Reserve Corps the Army Reserve. Perhaps most significantly, the 1952 legislation removed a caveat from the 1933 act by authorizing the president to order Guard units into federal service for fifteen days without declaration of a national emergency.

The debate that preceded the passage of the Armed Forces Reserve Act of 1952 saw the emergence of a relatively new phenomenon: competition between the Army's two reserve components, a competition that the Guard invariably won.[58] Initially, the National Guard Association opposed the bill because it "clearly favored the [Reserve] and did nothing for the Guard." The Guard argued that Congress should leave it alone; it could function effectively without new laws. Moreover, the Ready Reserve already existed in the National Guard. In response to the Guard, a Reserve general officer commented, "I had never realized that the main purpose of the federal government . . . was for the maintenance of the National Guard, and today was the first time I learned that the army was only an auxiliary of the National Guard."[59]

After the Korean War, the government conducted its traditional postwar military reorganization. The "New Look," as the nation's new military philosophy was dubbed, placed greater reliance on the reserves, and required the active Army, the Army National Guard, and the Army Reserve to "mesh smoothly and efficiently."[60]

Accordingly, the Army attempted to increase the training requirements for the reserves by requiring Guardsmen and Reservists to undergo six months of basic training. The National Guard managed to divine yet another Regular Army conspiracy to destroy it, and the National Guard Association fought the proposal. The highly respected Army chief of staff, General Maxwell D. Taylor, testified before Congress that six months was the absolute minimum amount of training time a reservist needed. The Guard responded by saying, "We know what we can do, and the Pentagon does not know. . . . [T]he offices of the Secretary of Defense and the Secretary of the Army are either bull-headed or trying to destroy the National Guard."[61] The controversy resulted in compromise legislation.

By the 1950s, the Army Reserve had also "developed into a potent political force and . . . the Reserve Officers Association had honed its political clout with the Congress."[62] During the Eisenhower administration, the Army Reserve succeeded in getting legislation passed over the administration's protest, including an overhaul of the reserve officer promotion system, budget increases, and a minimum paid drill strength.

During the late 1950s and early 1960s, the Guard's role in the civil rights crises in the south again highlighted the Guard's often conflicting state and federal roles. In 1957, Governor Orville Faubus of Arkansas called out the Arkansas Guard (as state troops) to block the entrance of black students into Central High School in Little Rock. President Eisenhower responded by federalizing the Arkansas Guard to enforce the Supreme Court's mandate to integrate. The Arkansas Guardsmen obeyed the president. This script was replayed in Alabama in 1963, when the Alabama Guard disregarded the orders of Governor George Wallace and responded instead to President Kennedy.

In 1961, the Army mobilized Guardsmen and Reservists in response to the Berlin crisis. After a short "honeymoon" period, the mobilized reservists became discontented with their situation, and in early 1962 some units staged demonstrations. This "unpredictable reaction" prompted the Army's decision not to send reserve soldiers overseas. The Berlin Crisis highlighted "the chronic problems of the Reserve Components—not enough equipment, not enough men and not enough training."[63]

Partly as a response to the Berlin Crisis fiasco, the Department of Defense proposed a spate of reserve component reorganization plans. In 1964, Secretary of Defense Robert S. McNamara announced a bold plan to merge the Guard and the Reserve. Unlike the Gray Board's recommendation, however, McNamara proposed to eliminate Army Reserve units by merging them into the Guard, leaving the Reserve as a pool of individuals. Naturally, the Guard supported McNamara's proposal; after all, it would have significantly increased the Guard's power. The Army Reserve and the Reserve Officers Association, of course, strongly opposed McNamara's plan, and their cause was abetted by the fact that 170 members of Congress were Reserve Officers Association members. Aided by McNamara's own lack of political savvy, Reserve opposition convinced Congress to kill the merger plan in 1965.

The war in Vietnam was yet another turning point for the reserves. One of the Army's more notable failures during the war was the failure to mobilize the reserve components. With a few insignificant exceptions, the conflict was fought by the Regular Army (of the 3.5 million soldiers in service during Vietnam, only twenty thousand were mobilized reservists).[64] Because the reserves, for the most part, stayed home, the Regular Army turned to civilians for its manpower, and consequently, the U.S. Army that fought the Vietnam War was a draftee army. (Incidentally, the reserve units that were deployed to Vietnam required extensive postmobilization training, as in past conflicts).[65]

The failure to mobilize had serious repercussions, both overseas and at home. In Vietnam, failure to mobilize led to a policy of individual replacement within Army units. Because they served two-year hitches, draftees were available to serve just over a year in the theater of operations. The nonmobilization policy thus created a constant turnover of personnel within units, contributing greatly to the lack of continuity and cohesion in those units.

Aside from its effect in Vietnam, the policy reverberated through the rest of the Army. Because soldiers served only a year in Vietnam before rotating home, U.S. Army Europe became a training ground for Military Assistance Command, Vietnam. A continuous stream of personnel flowed from Europe to Vietnam, and U.S. Army Europe units, therefore, suffered the same symptoms as units in Vietnam: a lack of stability caused by excessive personnel turnover. General Bruce Palmer, who served as the commander of U.S. Army Vietnam and later as vice chief of staff of the Army, has written that the failure to mobilize the reserve components contributed greatly to the emasculation of U.S. Army Europe, and indeed of the U.S. Army, as a fighting force.[66] In Europe, cooks were assigned as tankers; in the United States, second lieutenants commanded units with responsibility for nuclear weapons.

The decision not to mobilize them inevitably affected the reserves also. Because Vietnam was the Army's top priority and the rest of the Regular Army a distant second, the reserves were relegated to third-rate status. The Army provided the reserves with outmoded equipment and few resources (in fact, the reserves were forced to give up equipment to the active Army) and, in deeds if not words, neglected them. In sum, the nonmobilization decision "cast a stigma on the reserve forces that is yet to be completely erased."[67]

The nonmobilization policy's effect on hometown America was equally profound. The fact that Reserve and National Guard units were not mobilized contributed to civilian opposition to the war. Civilian communities had no hometown units overseas to support. Although many civilians knew individual soldiers who served in Vietnam, the sense of community involvement in the war effort undoubtedly would have been heightened had Guard and Reserve units been deployed. Indeed, because they were not mobilized, the Guard and Reserve came to be seen in a light quite different from the Regular Army. Because they understood the nonmobilization policy, many draft-age citizens—perhaps 70 percent of those who joined the reserves—enlisted in the Guard and Reserve expressly to avoid serving in Vietnam.[68]

The failure of the nonmobilization policy and the coming end of the draft spawned the total force concept in 1970. With the concept, the Department of Defense envisioned that in future conflicts the reserve components would fill the role that draftees filled during Vietnam: readily available military manpower with which to expand the active Army. Moreover, the total force concept would ensure that the failure of the Army's personnel policies during Vietnam—individual rotations and its concomitant effect on Army units—would not be repeated, because the concept envisioned the employment of the reserves as trained units.

With the end of the draft in 1973, the total force concept became the Total Force Policy. Under the Total Force Policy, "which integrate[s] the active duty, National Guard, and the other Reserve forces into [a] homogeneous whole," the Guard and the Reserve have assumed a greater role within the Army.[69] Many of the missions heretofore performed exclusively by the Regular Army were transferred to the reserve components. Thus the reserve components are no longer expected to be a third-string force to be mobilized and trained only as a last resort; rather, the Total Force Policy counts the Guard and Reserve among the vanguard of the Army, to be mobilized and to fight alongside the active component in the early stages of a conflict. Indeed, according to the chief of staff of the Army in 1986, the Army can employ a force of only five and one-half divisions without resort to mobilization of the reserve components.[70]

Accordingly, the reserves are an integral component of current war plans. War plans designed to fight the Soviets in Europe envisioned Army National Guard combat brigades *arriving in the combat theater* twenty-nine days after mobilization. More than 40 percent of Army forces deployed during the first thirty days of a European conflict under the Soviet scenario would have been reserve forces.[71] In effect, the Total Force Policy means that: (1) the reserve components are the chief source of additional combat strength for the active component; (2) the active component cannot respond adequately to an emergency of any significant size without mobilizing all or part of the reserve components; (3) readiness of reserve units must allow early mobilization and deployment with active units; and (4) the fundamental relationship of the reserve and active components is that of equal partners in peacetime preparations for war.[72]

But some things never change. In the 1980s, the Army deployed National Guard units, among others, to Honduras to train. Because they disagreed with the federal government's Central American policy, several governors protested the deployment of their troops, and two governors, under a provision of the law requiring their consent for federalization, refused. The move prompted Congress to enact legislation prohibiting governors from refusing to consent to the overseas deployment of National Guard troops "because of any objection to location, purpose, type, or schedule of such active duty."[73] The matter eventually ended up in the Supreme Court, which ruled against the governors and upheld the law's constitutionality.

Operation Desert Storm is the most recent chapter in the history of the Army reserve components. Its lessons are indeed important, so important that an entire chapter is devoted to that operation. For now, suffice it to say that the results of Desert Storm, both on the battlefield and in the political arena afterward, were entirely predictable.

If the history of the National Guard and Reserve is important, what has it taught us? Several trends flow through the 350-year history of the Army's reserves: (1) there has always existed a tension between the states and the Army with respect to controlling the National Guard; (2) the system is wedded to the geography of the country; (3) as local organizations, the external political power of the reserves, that which the reserves exercise collectively in the national arena, is formidable; (4) as local organizations, the reserves are also shaped by internal political factors, that is, within the reserve units themselves; (5) the reserves traditionally have been unprepared for war but eventually have managed to do the job (although a caveat should be added regarding support units in Desert Storm); (6) the citizen-soldier has habitually distrusted and disliked the regular, and vice versa; (7) the reserve components have tunnel vision—each is intensely parochial and protective of its own turf, so much so that they often lose sight of the forest for the trees and are quick to discern (or make up, as the case may be) "threats" from both the Regular Army and the other reserve component to their roles within the defense establishment; and (8) the Regular Army consistently, if not always consciously, has neglected the reserves as a matter of policy (a parochial attitude in its own right) and has failed to consider the very real political implications of its attempted reforms.

But the very fact that the reserves have a long and colorful history is important in itself, for the more an organization becomes steeped in tradition, the more resistant to change it becomes. Moreover, as an organization attains more prestige, it becomes increasingly loath to cede any of its power.

History, therefore, is an important starting point in any analysis of the reserves. It can teach us valuable lessons, and we must use it to prevent the repetition of past mistakes. Nonetheless, we cannot afford to remain wedded to the past; tradition for tradition's sake is the easy way out of tough problems. The challenges facing the reserve components in the twenty-first century bear critical examination, and although history cannot be ignored, the future, and not the past, must drive the result.

3. SYSTEMIC DISCONNECTS IN THE TOTAL ARMY CIRCUIT

> Any military system which tends to substitute separate machinery and set up competitive components with divergent objectives is harmful to service unity and thus to national security.
> —The Gray Board

THE Army's mission is to provide combat-ready units to the unified and specified commands.[1] Under the Total Force Policy, many of these Army units will be from the reserve components. Thus, because they must deploy quickly, many reserve units must be as ready in peacetime as active units. Although the Soviet threat has faded, the Army's reliance on the reserve components likely will increase as the active component shrinks, especially if the reserves, in light of increasing fiscal constraints, assume even more missions now performed by the active Army. The effect of the three-army system on the Army's ability to provide those reserve units is thus an issue of overriding importance.

That the readiness of the reserve components cannot measure up to that of the active component is well documented (*readiness* has three major components: training, personnel, and equipment). In the words of the former president of the now-disbanded U.S. Army Training Board, which conducted a detailed study of reserve component training, "No matter how well we do in peacetime training, the average [r]eserve unit will never be as combat-ready as its active-duty counterpart. Those who imply they will be as combat-ready are making a mistake."[2] The Congressional Budget Office in 1984 concluded that Army National Guard roundout battalions—the highest priority reserve component combat units—were rated 30 percent lower in training than comparable active battalions.[3]

But the problem is not merely that reserve readiness does not measure up to that of the active component. The problem is that the reserves cannot meet even lower, more "realistic" expectations because of the shortcomings inherent in our current reserve component structure and policies, products of 350 years of tradition and politics. This chapter begins the examination of the underlying systemic deficiencies contributing to these readiness shortcomings. Some of these deficiencies affect only one of the three facets of readiness; others span the readiness spectrum, magnifying their impact.

Readiness is directly (but not exclusively) related to funding. Reform of the budgetary process has done much to channel to the reserve components the resources required to improve readiness. Before Congress began bypassing the Department of the Army and appropriating procurement funds directly to the Army Reserve and the Army National Guard, the Army, at least according to Congress, skimmed funds from reserve component budgets and channeled them to the active component. In the words of Lieutenant General Herbert Temple, former chief of the National Guard Bureau, if it were not for Congress, the Guard would "be fighting with muskets."[4]

Obviously, though, the readiness of the reserves is still directly constrained by the availability of resources, which will no doubt continue to dwindle. At the end of fiscal year 1992, the Army National Guard had only 75 percent of its required equipment on hand, and the Army Reserve had but 66 percent.[5] The implications of these statistics are so clear they merit little discussion. A unit cannot go to war without its equipment.

Moreover, a unit cannot train without its equipment. Recognizing the obvious necessity of providing equipment for training, the Army has instituted the Minimum Essential Equipment for Training program, which is supposed to ensure that reserve component units have sufficient equipment with which to train. The program, however, has been implemented haphazardly,[6] and in any event, the program does not address how an Army National Guard tank battalion, for example, can train adequately when its tanks are kept at an active Army post several hours' travel time from its armories.

The equipment statistics do not indicate that much of the equipment reserve units do have is not compatible with the equipment of the active component units with which those reserve units will fight. A reserve component field artillery battalion with no secure communications equipment is absolutely useless as part of an active Army corps that relies on secure communications. Of course, there are not enough dollars for every active and reserve unit to be equipped with the newest equipment. Even if there were, equipment transitions take time, and wars do not necessarily wait until such transitions are complete.

The Department of Defense's sensible policy is that the units slated to fight first are equipped first. The problem with this policy is that reality often deviates from plans, and, as Desert Storm proved, those units programmed to deploy first often do not. Thus, after mobilization, the active component must scramble to ensure that the equipment of mobilized reserve component units can "interoperate" on the battlefield with active component equipment. Incompatible communications equipment spells disaster on the battlefield, as does a reserve maintenance unit assigned to support an active Army M-1 tank unit but equipped and trained to service the older M-60 model. Ultimately, the Army undoubtedly will cannibalize nondeploying active units to equip deploying reserve units, which could be disastrous if a conflict unexpectedly widens. Consider, for example, the consequences of cannibalizing the secure communications capability of a nuclear missile unit to equip a reserve howitzer unit in anticipation of a short conventional conflict. What happens when the next Saddam Hussein really does have something up his sleeve and the expected short conventional conflict becomes a long nuclear war?

Aside from the manifest impact of the federal budget, the reserve components are affected simply by having to share resources with the active component. The finest unit training opportunities in the Army, for example, are found at the Army's combat training centers, which have well-trained opposing forces, computerized feedback systems, and experienced trainers. The combat training centers include the National Training Center at Fort Irwin, California; the Joint Readiness Training Center at Fort Chaffee, Arkansas; and the Combat Maneuver Training Center at Hohenfels, Germany. The Joint Readiness Training Center is tailored to light forces; reserve component light forces consist of only one light division and two light roundout brigades. Thus, because the Combat Maneuver Training Center is obviously inaccessible to almost all of the reserves, the National Training Center is the only available combat training center for the majority of reserve combat units. The National Training Center, however, must serve the active component as a first priority, and opportunities for reserve units to avail themselves of training at the National Training Center accordingly are limited.

But more money alone cannot cure all of the reserve components' ills. They still have not been truly integrated with the active component. The Army National Guard, for example, maintains supply and personnel systems that are incompatible with those of the active component.[7] And, at the Department of the Army level, the Army's focus remains squarely on the active component. As a senior reserve officer noted:

> The Army chief of staff's three key deputies—the Deputy Chief of Staff for Operations [and plans] (DCSOPS), the Deputy Chief of Staff for Personnel (DCSPER), and the Deputy Chief of Staff for Logistics (DCSLOG)—have about 900 military officers working for them. These 900 officers are organized to provide the planning, resourcing, and policy work needed by the Army on a day-in, day-out basis. There are literally hundreds of defined managerial functions required of them, and most of these functions are just as pertinent to managing the 300,000 man Army Reserve and the 450,000 man Army National Guard as they are to managing the 780,000 man Active Army. Yet the overwhelming majority of functions performed by the DCSOPS, the DCSPER, and the DCSLOG are oriented on the Active Army. Relatively few Reserve problems, plans, or programs are worked to any depth.[8]

Active component assignment and promotion policies highlight the Army's focus on that component. The Army typically does not reward the active officers it assigns to reserve component support duty, and typically does not assign its best officers to that duty. Not surprisingly, therefore, active officers seek to avoid service in reserve-support positions.

Aside from the failure of the Total Force Policy to bring together the active and reserve components, the policy also has not eliminated the rivalry between the two reserve components. Although the competition between the Army Reserve and the Army National Guard certainly is neither as intense nor as overt as it once was, it is nonetheless present. Both reserve components compete for resources from the same pot, and each is anxious to boost its own status. A long-range goal of the Army Reserve, for example, is to add combat units to its ranks[9]—and such units can be added only at the expense of the Guard.

Several factors cause the components to remain separate. Our trifurcated structure itself is perhaps the most noticeable detriment to the Army's readiness. In the words of James L. Lacy, former special assistant to the assistant secretary of defense for manpower, reserve affairs, and logistics (from which the position of assistant secretary of defense [reserve affairs] evolved), "Selected Reserve organization remains a mind-boggling array of separate bureaucratic compartments."[10]

Army National Guard and Army Reserve units, for the most part, are controlled not by tactical headquarters but by administrative organizations. Although they are charged with the mission of preparing troops for combat, these headquarters will not employ in combat the forces they train in peacetime. This system contributes greatly to ineffective reserve unit training.

Both active and reserve units are supposed to prepare "mission essential task lists" that, based on the premise that even an active unit cannot train on every conceivable task, narrow and focus the scope of unit training according to a unit's wartime mission. The mission essential task list, therefore, is the starting point for any collective training program.

Both the Army's inspector general and its Reserve Component Training Strategy Task Force found that some reserve component units do not prepare mission essential task lists and others prepare them improperly, either because commanders lack experience or because guidance from higher headquarters is vague.[11] In either case, the mission essential task list problem points out the larger systemic deficiency. Guidance from which to develop a mission essential task list is often vague because the headquarters issuing the guidance are not in the warfighting chain of command. They are thus incapable of providing training guidance based on their own wartime missions, nor can they translate guidance from the CAPSTONE headquarters adequately. Moreover, because they are concerned with administrative matters, these headquarters have neither the systemic incentive nor the expertise to review faulty mission essential task lists and provide the appropriate direction to correct them.

These administrative organizations have a penchant for generating multiple and unnecessary requirements that are often unrelated to combat readiness. The Reserve Component Training Strategy Task Force reported that the average reserve company commander is burdened with at least 115 administrative requirements annually, many of which are duplicative; according to the Army, a reserve component unit spends one-half of its thirty-nine days of annual scheduled unit training time complying with administrative requirements.[12] In the words of the Reserve Forces Policy Board, administrative emphasis creates a reserve chain of command "that is insensitive to the negative impacts caused by requirements it imposes on subordinate units. This situation tends to reward administrative prowess rather than effective training. In many cases, unit commanders are more comfortable concentrating on administration and other [readiness] detractors because results are measurable."[13]

Although these systemic problems exist in both the Army Guard and the Army Reserve, they are more apparent in the Army National Guard. The Guard is in reality fifty-four separate forces, each responsible to a different commander in chief. Since 1933, the Guard's statutory structure has changed little; in peacetime, the Guard still is commanded by the states. The Army's legal authority over the Army National Guard is limited to monitoring and advising. *The Army has absolutely no legal command authority over the Army National Guard when it is in state service*—that is, almost always.

The Guard, when in state service, is commanded by the governors through their adjutants general. Although the adjutants general normally have risen through the Guard ranks, they are state officials in every case. They are appointed by the governors (except South Carolina's, which is popularly elected) and therefore owe their first loyalty to the states, not to the U.S. Army. Federal funding is the only systemic incentive for these officials to do the Army's bidding.

Federal funding, however, is an inadequate incentive. Federal funds are too big a stick to use for small carrots. To deny federal funds to the Guard, the Army must be prepared to wind up in the federal courts, which is exactly what occurred when the National Guard Bureau threatened to deny funds to states whose governors protested the deployment of their Army National Guard units to Honduras.[14] Litigation is a steep price to pay, and the Army therefore conveniently overlooks many of the Army National Guard's lesser deviations from Army guidance. But these allegedly minor deviations add up.

The effect of the state-federal dichotomy is profound. There are currently ten divisions in the Army National Guard (this number is scheduled to decrease), or roughly 42 percent of the Total Army's combat divisions. Moreover, under the roundout program, the Guard provides a significant portion of the combat power for active component divisions. As of 1992, Army National Guard brigades round out the 1st Infantry Division (Mechanized), the 2d Infantry Division, the 4th Infantry Division (Mechanized), the 5th Infantry Division (Mechanized), and the 10th Mountain Division (Light Infantry), all of which are active component divisions (additionally, an Army Reserve brigade rounds out the 6th Infantry Division [Light]). Two other active divisions, the 1st Cavalry Division and the 24th Infantry Division (Mechanized), have an Army National Guard "roundup" brigade, a fourth maneuver brigade in addition to their three active component maneuver brigades. In all, eight of the fourteen active component divisions currently have reserve component brigades programmed into their structure.[15] The Guard thus provides a substantial amount of the Army's combat power, and much of that combat capability is ostensibly an integral part of active component divisions.

But the Guard system of state control, the vestige of the colonial militia and of the militia clauses the framers of the Constitution drafted to preclude reliance on a professional standing army, subverts the Army's ability to accomplish its mission. The system ensures that Guard units, especially those that include multistate subordinate elements, will not be combat ready when the Army needs them. State control has two major effects: first, it ensures that the Army's control over its Army National Guard units is illusory; and second, it ensures that Army National Guard commanders cannot control even their own units.

Eight of the ten Army National Guard divisions consist of

units from more than one state. In the 35th Infantry Division (Mechanized), for example, the division headquarters and the 2d Brigade are part of the Kansas Army National Guard; the 1st Brigade is part of the Nebraska Army National Guard; the 3d Brigade is part of the Kentucky Army National Guard; and the division support command headquarters is part of the Missouri Army National Guard and "controls" subordinate battalions from Kansas, Kentucky, Missouri, and Nebraska. In peacetime (i.e., almost always), although the division commander communicates with all of the elements of his division, his authority to command the division is limited to Kansas Army National Guard units.

The practical effect of this system can best be illustrated by relating some experiences with the Guard unit of which I was a member, the 29th Infantry Division (Light). The 29th Infantry Division is a unit with a proud heritage. The division fought commendably during the Second World War, landing on the Normandy beaches on D-Day. Today the division consists of subordinate units from both the Maryland and Virginia Army National Guard.

Several events surrounding one of the division's recent annual training periods provide clear examples of the deleterious effects of peacetime state control of the Guard. The commanding general of First Army, the CONUSA with supervisory responsibility for the division, issued training guidance indicating that he intended that units for which First Army was responsible spend nine consecutive days of their annual training periods in a completely "tactical" environment (this training guidance reiterated a Forces Command regulation). An officer in an active light infantry division would interpret this guidance to mean that units were to train continually for nine days, with no "administrative" breaks. In other words, the First Army commander intended reserve component units to live, eat, and sleep as they would be required to do in combat. This guidance reflected the professional judgment of a career Regular Army general officer, an officer who himself had commanded an active component light division.

The adjutant general of Virginia, however, had other ideas. Concerned that attrition in the Virginia Guard was increasing because the Guard had narrowed its focus to concentrate on training to the exclusion of other, more "fun" events that enhance camaraderie in a different way, the adjutant general decreed that although units would stay "in the field" for nine days, they would establish administrative bivouacs at night, complete with aligned pup tents. Moreover, the soldiers would be transported from the field to garrison occasionally during the nine-day period so that they would have the opportunity to take a shower and to drink a beer at the post exchange. Not only did the adjutant general's orders contravene the guidance of the Regular Army officer charged with supervising the training of the 29th Infantry Division (Light), it undermined the Army's concept for the light infantry divisions, which stresses that thorough, rigorous training is to be the top priority and that training detractors (and how else can "PX runs" be classified?) must be eliminated.[16]

This incident aptly demonstrates the ability of a state official to subvert the intent of the U.S. Army. Of course, the CONUSA commander could have made an issue of the affair and threatened to have federal funds withheld, but the Army must pick its battles judiciously. I cannot state unequivocally that this incident is typical of the Army Guard, but it certainly proves the point: the Army must depend on the states to make Army National Guard units combat ready and can only advise the states concerning how to accomplish that mission. And the states can—and at least sometimes do—fail the Army.

A second incident illustrates the lack of control the division commander has over his own subordinate units. Before the commencement of annual training, the division commander (a Virginia Guardsman) issued his training guidance to the division. This guidance was, of course, disseminated to both the Virginia and the Maryland units. The Virginia units followed the guidance. The 3d Brigade, however, a Maryland unit, decided (or perhaps was directed by the Maryland adjutant general) to perform a "firepower demonstration," a range display of the brigade's weapons systems for politicians and other Maryland dignitaries. This exercise involved a significant portion of the brigade and consumed an inordinate amount of staff planning time and effort (the deputy brigade commander did nothing during the annual training period but plan and supervise the firepower demonstration). This event bore little relation to the division's readiness or to the division commander's overall training guidance. Moreover, the division actually supported the brigade in staging this extravaganza because the division commander and his staff realized they had no other choice.

As this incident demonstrates, even within their own units Army National Guard commanders are hamstrung by state control. Unit commanders are absolutely powerless to implement their own guidance to subordinate units from different states. Indeed, a Guard commander from one state has even less power over subordinate units from another state than the Army has over the Guard as a whole; at least the Army has the power to cut off funds. The 29th Infantry Division, however, has no fiscal authority over its 3d Brigade because the brigade is funded through the state of Maryland, not through the division headquarters in Virginia. The words of the Gray Board, the committee formed in 1948 to report to the secretary of defense on the state of reserve forces, still ring true: "In peacetime the officers of one State can command troops of another State only through indulgence, courtesy, or comity, even though organizations may overlap state lines." [17]

In practice, multistate Guard units are commanded by consensus among the nominal unit commanders, the respective state adjutants general, the CONUSAs, and the National Guard Bureau. Furthermore, through the CAPSTONE program, even the gaining wartime commanders, by providing training guidance based on the programmed wartime missions of Army National Guard units, supposedly have a hand in telling those units what to do.

To govern the 29th Infantry Division, for example, the Maryland and Virginia adjutants general and the chief of the National Guard Bureau signed a formal "memorandum of understanding." This memorandum created a "division council" comprising the two adjutants general, their principal deputies, and the division commander. The council selects the division commander, rotating that position between the two states. Each adjutant general appoints one of the two assistant division commanders (brigadier [one-star] generals' positions), who serve at the pleasure of the respective adjutants general; in other words, the division commander has virtually no formal authority over his two general officer assistants. The council approves the division's training plan after it has received previous approval from each state. Finally, should the entire division be mobilized, the Maryland and Virginia state area commands retain responsibility for mobilizing the division's elements within their states. When all is said and done, the power brokers with the biggest vote are the adjutants general, who, of all the interested parties, have the least incentive to ensure that their state troops are ready for their federal missions, and may even be at cross purposes with one another.

The Gray Board summed up the negative effects of state control: Dual control "produce[s] a constant turmoil of bickering, recrimination, factionalism and stalemate. Constructive planning cannot be directed toward national security as the sole objective. It is constantly running into political or partisan stalemates. . . . A divided program results, which compromises security and satisfies neither party. . . . The present organization of the National Guard does not repose authority where basic responsibilities rest. None of the State and Territorial Governors has any direct responsibility for the national security and none is answerable to the national government."[18]

This system violates the basic principle of Army training, which states that a unit should train as it will fight.[19] More simply, the system violates good military sense. Under the current system, the active component commander who will receive a Guard unit upon mobilization is charged with responsibility for employing that unit in wartime but has virtually no hand in the preparation of that unit for combat. The system also violates fiscal common sense, because it lets the adjutants general, state political appointees, directly control federal funds to train a substantial proportion of the Army's combat capability, and it pays Army National Guard soldiers with federal funds to perform state duty.[20] The state-federal structure is akin to Ford contracting with General Motors to produce Ford cars at a price higher than Ford itself could produce them.

The systemic flaws that inhibit the readiness of the Army Reserve are more subtle than those of the Guard, but they are nonetheless present. Like the Guard, the Army Reserve is a world unto itself, linked with the active component only at the highest levels.

Although Forces Command actually commands Army Reserve units in the continental United States, that headquarters is so far removed from Army Reserve units that those units suffer many of the same symptoms as their Army National Guard counterparts. The ARCOMS are in many ways analogous to the state headquarters that control the Guard. They are two-star administrative headquarters with no wartime missions (at least with regard to their subordinate units; some ARCOMS have missions to run active Army posts in the event the post headquarters personnel deploy, whereas other ARCOMS are charged with reconstituting the Army Reserve in the event of full or total mobilization). Like Army National Guard units, the ARCOMS' subordinate units have no formal peacetime command relationship with their wartime gaining commands.

Even the general officer commands, although a step closer to their gaining commands, are in the same predicament. The general officer commands report directly to USARC (or they will when USARC assumes full control of Reserve units from the CONUSAS). Like the ARCOMS, USARC is an administrative, not a warfighting, command. It is responsible not for providing Army Reserve troops to the combatant commands, but for providing Army Reserve troops to the active component. In this regard, USARC is no improvement over the arrangement that existed (and, during the transition period, continues to exist) when the CONUSAS commanded Army Reserve units; USARC, in fact, simply functions as a large ARCOM.

Like the states as they relate to the Army National Guard, Army Reserve administrative headquarters have no battlefield stake in the results of their peacetime efforts. If their troops are ill prepared, the ARCOMS will not fail on the battlefield. In contrast to a warfighting commander whose incentive to train his subordinate units is to ensure that his own unit is successful in combat, the only systemic incentive ARCOM commanders have for ensuring the readiness of their subordinate units is their own ambition and professional pride.

This is not to say that this incentive is neither strong nor valid; it certainly is both. But the lack of a more substantial stake in their subordinate units' readiness provides an incentive for administrative headquarters to accept blindly their subordinates' assessments of their readiness, or even to inflate readiness estimates. The incentive of these commanders is one of peacetime success rather than of wartime success—to look good rather than to be good. Again, in civilian terms, this situation is like contracting for work but telling the contractor that he will make no profit. There is no incentive to ensure that the work will be of high quality.

Even when USARC fully displaces the command functions of the CONUSAS this fact will not change. USARC too has strictly a peacetime incentive. Created because the Army Reserve was allegedly a "second class citizen" within the Army, USARC may give the Reserve more clout in the Pentagon, but, as a purely administrative headquarters, it assuredly will have a considerably lesser effect on the readiness of its subordinate units.

What about CAPSTONE, the tie that allegedly binds all these disparities together? The answer is simple: CAPSTONE does not work and it never has. Both the active component and the reserve components share the blame.

CAPSTONE is a peripheral player in the life of a reserve unit. The reserve component unit hears occasionally from its wartime gaining command, and it accomplishes the wartime commander's guidance insofar as possible. But whenever an entity must respond to more than one distinct boss, there are bound to be conflicts. These conflicts may be comparatively minor; in a system in which every boss must contribute to accomplish the mission, however, even minor conflicts are significant. Ultimately, the boss with the most clout commands the greatest respect, and satisfying that boss becomes the top priority. In the current reserve system, that person is the state adjutant general in the case of the Army National Guard, and the ARCOM or USARC commander in the case of the Army Reserve. None of these commanders has a warfighting mission.

From the active Army perspective, CAPSTONE is also a minor part of daily life. Active component commanders focus on their own active units. The reserve component units affiliated with those active component units through CAPSTONE may be slated to become part of the active units upon mobilization, but because reserve units are not part of those active units on a daily basis in peacetime, they do not garner as great a share of the wartime gaining commander's attention as the active component units the wartime commander "owns." This is, of course, completely predictable; without a formal command link between the reserve unit and its wartime gaining commander, the active commander has no real incentive to make the time to oversee the preparation of his CAPSTONE reserve component units for war. Moreover, he cannot watch those units from the window of his headquarters, like he can his active component units. And, ultimately, the active commander who does attempt to monitor his CAPSTONE reserve component units is bound to be frustrated by his inability to ensure that his guidance is implemented.

On a more practical level, consider how a directed training association is often implemented. An active component division that has a directed training association with a reserve component separate brigade generally is tasked with supporting and evaluating the brigade's annual training period. As a rule, the active component division staff does not consider who in the division would be best qualified to evaluate the annual training. The mission is simply another support "tasking" to be accomplished. The division staff decides how many evaluators and support personnel are needed. It then "tasks" the subordinate brigade that is pulling its duty as the post support brigade to provide those people. (Most active component Forces Command divisions rotate their subordinate units among a "mission" cycle, when the unit is at its highest readiness for deployment; a training cycle, when the brigade trains intensely; and a support cycle, when the brigade supports the training of the other two brigades and performs the housekeeping chores attendant to the small cities that Army posts are—cutting grass, picking up trash, raising the flag in front of the post headquarters, etc.)

The brigade that receives the tasking from the division headquarters considers it a significant training detractor—which, undeniably, it is—and attempts to accomplish the evaluation and support mission with as few resources as possible. For example, the active component brigade is required to write an evaluation report for the reserve component brigade headquarters company. The headquarters company comprises several disparate elements: the brigade staff, which is itself split tactically between the main command post, the tactical command post, and the brigade support area; the communications platoon; the mess section; and the company headquarters, to name a few.

Despite this disparity, because only one evaluation report is required for the company, the active brigade sends only one evaluator. Of course, the evaluator cannot possibly observe each of the elements of the company in enough detail to evaluate them thoroughly, nor can he possibly be qualified to evaluate every element of the company.[21] Moreover, the evaluator often "sugar coats" his evaluation.[22] In the final analysis, the reserve unit fails to receive the intended benefits of the directed training association. The brigade headquarters company does not receive a thorough evaluation, nor does it benefit from the informal advice that it would receive were there as many qualified evaluators as there ought to be.

The flaws in the evaluation system manifest themselves in the readiness reporting system. Evaluation reports are often inaccurate.[23] Thus, the Army really does not know the true readiness of the reserve components. In the words of one commentator, the Army "simply pencil[s] in or pencil[s] out whatever degree of readiness is supportive of the bureaucratic and budgetary interests of the moment."[24]

Moreover, unit status reports, which report unit readiness, are transmitted through the peacetime administrative chain of command, a systemic flaw that further separates reserve units from their wartime gaining commands. Consequently, even if an active component gaining commander did everything envisioned by CAPSTONE, he would not routinely receive the reported results of his efforts. The wartime gaining commander does not know the reported status of his CAPSTONE units until those units show up on his doorstep after mobilization.

Like reserve component units, the Individual Ready Reserve, as a whole, also falls short of expectations. The Individual Ready Reserve, in Lacy's words, "may best be described as a telephone directory of people who have been in active or Selected Reserve service and who are still liable to call-up in emergency. A cohesive and credible force it is not."[25]

The Individual Ready Reserve is often used as a dumping ground by Selected Reserve commanders to rid their units of substandard performers; administratively, transferring a soldier from the Selected Reserve to the Individual Ready Reserve is much easier than separating that soldier from the Army altogether. At the end of fiscal year 1986, 44 percent of the enlisted members of the

Individual Ready Reserve had been transferred to the Individual Ready Reserve in lieu of discharge.²⁶ Additionally, concerned about the shortage of pretrained manpower, Congress directed the Army in 1978 to retain in the Individual Ready Reserve soldiers who, for administrative reasons, were ineligible for reenlistment in the active component, even if those soldiers received general discharges from the active component "under honorable conditions" (this type of discharge is not an honorable discharge). Thus, many of those in the Individual Ready Reserve telephone directory are soldiers that active component and Selected Reserve units would be better off without. They are hardly "mobilization assets."

In 1987 the Department of Defense instituted a one-day annual muster for Individual Ready Reserve soldiers. This "day," however, is more accurately an hour. Although Army policy is that the annual Individual Ready Reserve "screen" should include the status of the Individual Ready Reservist's physical condition and skill proficiency degradation,²⁷ in reality the Individual Ready Reserve soldier reports to an Army recruiting station and sits in an office while a recruiter fills out a form verifying the soldier's home address and other rudimentary personnel data. Moreover, in fiscal year 1990, only 13.1 percent of Individual Ready Reservists were screened.²⁸

The Army counts as "trained" only those Individual Ready Reserve soldiers who have transferred from the active component or the Selected Reserve within one year, those who have been deemed trained through the Individual Ready Reserve screen, and those who have performed active duty within the past year.²⁹ By the Army's own calculation, there were only 43,000 trained Individual Ready Reservists (of a total of 293,000) as of September 30, 1988.³⁰ In actuality, the number of trained Individual Ready Reserve soldiers is probably even more minuscule; of the three categories, only most of those soldiers who have separated from the active component or Selected Reserve within one year, as well as those whose civilian occupations parallel their military specialties, should be so considered. As far as verifying the training status of soldiers whose civilian occupations do not mirror their military specialties, the annual Individual Ready Reserve screen is completely ineffective.

Training for Individual Ready Reserve soldiers is a low Army priority, and it will get lower as budgets get tighter. In fiscal year 1988, only 4 percent of the Individual Ready Reserve received training. Moreover, the Individual Ready Reserve training that does occur does not support mobilization priorities. As the General Accounting Office has reported:

> The IRR training program does not focus funds on skills that require refresher training and are most needed in the early days of conflict. Instead, the Army's first priority for [Individual Ready Reserve] training is professional development education. In addition, mobilization training funds are used (1) for non-training purposes such as participation in a marksmanship program, (2) for multiple training tours to some individuals even though most soldiers received no training, and (3) for training in military specialties not required in the first 30 days of mobilization.[31]

For those Individual Ready Reserve soldiers who do receive mobilization-related training, active-duty tours normally consist of two-week assignments to active component or Selected Reserve units. Whether the Individual Ready Reserve soldiers actually receive meaningful training during those two weeks depends on what the unit is doing and, for noncommissioned officers and officers at least, the degree of responsibility the unit leadership decides to give them.

Because the Individual Ready Reserve consists largely of soldiers who separate from the active component with remaining service obligations and who choose not to serve in the Selected Reserve, the size of the Individual Ready Reserve relates directly to the size of the active component. As the active component shrinks, so will the pipeline to the Individual Ready Reserve eventually shrink, although there will be a short-term influx provided by the reduction of the active force and by the institution of the eight-year service obligation. During full mobilization, the Army will rely on 115,000 Individual Ready Reserve personnel in the first thirty days after mobilization just to bring units up to strength, discounting those needed to replace casualties (remember that the Army estimated in 1988 that it had only forty-three thousand trained Individual Ready Reserve soldiers).[32] In a 1987 test, however, only 53 percent of Individual Ready Reservists ordered to active duty actually reported.[33] Thus not only will the Individual Ready Reserve not have enough *trained* manpower to fill deploying units adequately, it may not have enough manpower at all. (Granted, the Individual Ready Reserve reporting rate for Desert Storm was about 75 percent;[34] as explained in chapter 5, however, the Desert Storm Individual Ready Reserve mobilization was not a true test of the system.)[35]

Beyond sheer numbers, however, the Individual Ready Reserve has yet another shortcoming. Even if the Army had enough soldiers in the Individual Ready Reserve, and even if all of those soldiers were trained in their specialties, there is still the issue of skill mismatches. The Individual Ready Reserve does not contain the right mix of people with the right skills to support the Army upon full mobilization.[36] According to one author, only one-third of the Army's Individual Ready Reservists are trained in combat and medical skills, but more than two-thirds of those needed would require such skills.[37] Thus, even some of the trained soldiers in the Individual Ready Reserve may not be mobilization assets because they are trained in skills the Army may not need immediately.

The National Defense Act of 1933 and the Armed Forces Reserve Act of 1952 were the last substantive reforms of the reserve components. Those laws created the National Guard of the United States and the Ready Reserve. Although the reorganization of the reserves may have seemed promising in 1933 and 1952, we still have a reserve structure that is archaic and inefficient. And although Congress has thrown more money at the reserves, the underlying problems remain.

The organization of the Selected Reserve and the policies regarding the Individual Ready Reserve are more than just antiquated and inefficient. They are harmful and detrimental to the sine qua non of the U.S. Army: readiness for combat. Control of the reserves by the states and other administrative headquarters, lack of involvement of wartime commanders in peacetime preparation for war, the hollowness of the Individual Ready Reserve—all detract from readiness. If Congress and the Army are serious about being ready to fight the next war, they must address these problems.

4. GEOGRAPHY, TIME, AND OTHER READINESS DETRACTORS

> Such a degree of dependence [on the reserves] would not be a source of concern if the reserve components came close to matching their active-duty counterparts in capability and readiness. But they do not.
> —Martin Binkin and William W. Kaufmann

THE reserves' problems run deeper than organization and structure. Indeed, many of the impediments to reserve readiness can be traced back two centuries.

The first colonists brought the citizen-soldier concept with them to America from England in the seventeenth century. A key feature of the citizen-soldier force in colonial times was its geographic organization, a throwback to feudal England. This characteristic continues to pervade the reserve components today, and geography is a major impediment to reserve unit training, a key ingredient of readiness.

Most reserve component units are located significant distances from suitable collective training areas. On the average, a reserve unit must travel 40.1 miles to the nearest local training area, where its soldiers usually can perform small unit training without firing live ammunition. To reach a major training area where it can conduct the same training generally available to active component units at their home posts, the average reserve component unit must travel 154.2 miles. The average reserve unit must travel 65.7 miles to the nearest rifle range just for soldiers to qualify on their individual weapons.[1]

Whereas active Army units can walk out of their barracks to their motor pools to reach their tanks, howitzers, or combat engineer vehicles, reserve units' major systems are stored at mobilization and training equipment sites or equipment concentration sites. Far from being able to walk outside of their armories to maintain their heavy equipment, reserve units must travel an average of 128.5 miles just to get to it.[2]

Time is another major factor that significantly affects reserve component unit training. Even on the surface, the effect of time is obvious: a reserve component unit that trains together thirty-nine days annually cannot be expected to measure up to a similar active component unit that trains continuously throughout the year. Granted, active component units do not train every day. Undeniably, though, they train more than do reserve units, planning for approximately two hundred days of training annually.[3]

The dedication of reserve component soldiers is perhaps unmatched, even by active component soldiers. Many reserve soldiers devote much more time individually to their units—without pay and on top of full-time civilian careers—than the thirty-nine days of scheduled unit training. Nonetheless, except in special circumstances, the time these soldiers dedicate to the Army is generally spent planning and administering, not training. *Unit training*, as a rule, is scheduled for only the designated thirty-nine days.

When combined, the effects of geography and time are magnified many times. In contrast to the Air Force reserve components, which can train at the local airport or Air National Guard base, many Army units need large areas to train collectively. A mechanized infantry battalion, a field artillery battery, or a signal unit all require significant maneuver space to train with their primary tactical systems. In addition, combat units need ranges and impact areas to fire their units' weapons systems.

Such areas are obviously available only on military installations, which, as explained above, are not readily accessible to reserve component units. Although the Army is implementing a concept of regional training centers, those centers are oriented primarily toward individual combat service support training that cannot be conducted effectively at home station.[4] The concept does not address the problem of effective collective training, especially in combat units.

Thus, many reserve component units, especially combat units, can accomplish little significant collective training during a weekend drill, simply because thirty-six hours (Saturday morning through Sunday evening) is not enough time to organize a unit that has not been together for a month, issue equipment, travel to a training area, train, return to home station, maintain the equipment, and turn it back in. A reserve component officer writing in *Infantry,* a professional journal, proposed a weekend training schedule that would optimize available training time; of the thirty-four total weekend hours in that schedule, only eighteen are devoted to training (assuming the unit's soldiers do not sleep).[5] The combined effect of time and geography on training is therefore even more detrimental than it first appears. Lack of meaningful weekend training affects readiness in other ways as well: it is a prime contributor to reserve attrition problems.[6]

The reserve components' inability to conduct adequate collective training on weekends means that they can conduct such training only during their annual training periods. But even some of that time, at least one day at the beginning and two days at the end, is necessary to organize, maintain equipment, and travel to and from the annual training site. The bottom line is that there are an absolute maximum of twelve prime training days annually for a reserve unit, and the average unit conducts only eleven days of effective field training during a two-week annual training period.[7]

As if not having enough time to train were not detrimental enough, reserve component units often use the time they have inefficiently. Many reserve brigades and divisions spend part of their annual training periods conducting brigade- and division-level exercises, the type of training that even the active Army does not conduct routinely. This large-scale training is counterproductive. Higher-level exercises in the reserve components are conducted at the expense of essential individual and small unit training. Because geography and time prevent individuals and smaller units from reaching sufficient levels of training, the basic building blocks necessary to conduct these types of exercises are absent. Although the training is helpful for brigade and division staffs, subordinate units become mere tools to assist in that staff training. Smaller units and individual soldiers get little value out of such exercises; for them, it is training time wasted.[8]

Many of the factors that adversely affect collective training also inhibit individual training. Even individual training, which reserve units can conduct on weekends, is a much greater challenge in the reserve components than in the active component. As a journalist (who also happens to be a retired Marine Corps general officer) has written, "Only reservists whose wartime missions would parallel their civilian jobs . . . can easily maintain the level of combat-readiness expected of them."[9]

Statistics bear out this assertion. In 1987, for example, only 65 percent of reserve component soldiers tested passed their skill qualification tests, which evaluate soldiers' proficiency in their military occupational specialties.[10] Combined with the fact that 21 percent of the soldiers in Selected Reserve units fill positions for which they are not qualified,[11] the skill qualification test statistic is unsettling: it means that less than half of all soldiers in reserve component units are proficient in performing their military duties.[12]

The individual training challenge is especially tough in the Army Reserve. Forty-six percent of the Army Reserve force structure is combat service support, compared with 16 percent for the active component and 17 percent for the Army National Guard.[13] Considering that the Army Reserve relies on the active component for a significant portion of its manpower, the numbers just don't add up: there are not nearly enough combat service support soldiers who separate from the active Army to provide an adequate number of qualified soldiers to the Reserve. The Army Reserve, therefore, must retrain many of the soldiers it receives from the active component.

Even more troubling are the results of the common task test in the reserve components. The common task test examines those skills that every soldier, regardless of military occupational specialty, is required to possess to survive on the battlefield, such as wearing a chemical protective mask, providing basic first aid, and recognizing enemy tanks. Thirty-two percent of Army National Guard soldiers tested (remember that the Army National Guard is predominantly a combat force) failed the recognition of friendly or enemy armored vehicles test, as compared with 11 percent for the active component. Twenty-eight percent of Army National Guard soldiers failed the personal chemical decontamination test; 5 percent of active Army soldiers failed.[14]

Poorly trained individual soldiers directly and adversely affect the quality and effectiveness of unit training. The Reserve

Forces Policy Board has stated that some commanders are incapable of planning and conducting meaningful training.[15] The General Accounting Office has found that noncommissioned officers lack the basic battlefield survival skills necessary to integrate combat realism into collective training.[16] The quality of unit training is totally dependent on the skill of the men and women who plan and execute it. Without trained reserve leaders there will be no trained reserve units.

As it affects training, so too does geography have personnel readiness implications. Reserve component units are composed of soldiers from the same communities. Depending on the size of the communities, the local National Guard armory or Army Reserve center may be home to a battalion or only part of a company. In more rural areas, a battalion may be scattered throughout five separate communities. In fact, the average reserve battalion is dispersed over a 150-mile area; some are dispersed over areas as large as 300 miles.[17]

In contrast to the active component, which can recruit from one large national manpower pool, the Guard and Reserve have discrete and limited pools—their surrounding communities—from which to man their units. Proponents of this system cite several perceived advantages. Civilians tend to remain in the community, whereas active personnel transfer in and out of units every few years. Local reserve component units, so the argument goes, tend to be more cohesive than active Army units because of supposedly lower personnel turnover and because of the familiarity born of close community ties among the soldiers.

But the average time soldiers remain in the same unit may not be as long as some proponents of the system aver. In many units, the shibboleth that soldiers remain in the unit for thirty years is simply untrue. As urbanization spreads, reserve units, dependent on population density for recruiting pools, also must move to the urban areas. The urban civilian population is just as peripatetic as the active Army. Moreover, even in units whose personnel tend to remain for years, personnel turbulence caused by internal reassignments, promotions, and attrition are facts of life, just as they are in the active component.

Thus, the claim that personnel turnover is lower in reserve

component units than in active Army units is unfounded. Indeed, the annual turnover rate for junior enlisted soldiers in reserve units is 50 percent.[18] The effect of that turnover in the reserve components is considerably more severe than in the active component because reserve units must recruit to replace their personnel losses; unlike the active component, they cannot simply request the Army to assign a replacement. Such personnel turbulence adversely affects unit effectiveness and morale.[19]

Even if turnover in reserve units were nil, such constancy could well be a readiness inhibitor in itself. Units with no infusion of new blood stagnate. They are slow to change, and they resist fresh ideas. A common refrain in units whose soldiers have been together too long and whose leaders have never served in other units is "that's the way we've always done it here." Such attitudes are not conducive to producing combat-ready units. On the contrary, the result of such unity is stable, cohesive, inexperienced, and untrained units. In the words of James Webb, former assistant secretary of defense for reserve affairs and former Marine combat infantryman, "That kind of camaraderie is no replacement for training. All it means is they'll help each other a lot in the POW camp."[20]

Even if the cohesiveness argument were true, however, geography negatively affects personnel readiness in other ways. Force structure decisions are based on the requirements of the Army as a whole, and these decisions have a disproportionate impact on the reserve components. When the Army decides to create a new active component signal battalion, it forms that battalion from soldiers throughout the active Army who are qualified in their military occupational specialties. On the other hand, when the Army decides to transfer the mission of a signal battalion from the active component to the reserves, it must form that reserve component signal battalion from reserve component soldiers in a particular geographic area.

Thus the Army has little choice but to convert an existing reserve unit into a signal battalion. When a reserve tank battalion becomes a signal battalion, the battalion does not receive soldiers who are qualified in their specialties; it must turn its tankers into communicators. Therefore, although soldiers may remain in the local reserve unit for long periods, the reserve unit itself may change as the Army's force structure evolves, and when this happens the reserve unit is stuck with its demographics.

Demographics may disrupt the reserve components even more severely in the near future. Former Secretary of the Army Michael P.W. Stone stated that "there are significant force structure problems related to local demographics, social and employment conditions."[21] As the nation's population shifts from the northeast to the west and south, a geography-based reserve system will have to follow the population.[22] Thus units on the East Coast will have to be disbanded, and units in the West will have to be established. The readiness implications of such actions are self-evident.

Although certainly not as severe, the effect is much the same when a unit receives new equipment. When a reserve tank battalion equipped with M-60 tanks receives newer M-1s, the unit undergoes a learning curve that is considerably longer, in terms of calendar time, than the adaptation period of an active Army unit that undergoes the identical transition. Nor are such transitions within the reserve components isolated occurrences. In fiscal year 1986, for example, nearly one of every twenty reserve units underwent some type of conversion. That year 122 reserve component units were activated, 18 were inactivated, and 233 underwent major conversions.[23]

The geographic hometown reserve system has another negative effect on personnel readiness. The active component is a major source of trained manpower for the reserve components. But when a trained engineer who desires to continue his military service leaves active duty and returns home to join the local reserve unit, his training as an engineer is wasted if his hometown unit is a field artillery battery. The engineer will have to be retrained as an artilleryman. Each year, about one-quarter of the enlisted personnel who enter the Army National Guard are soldiers with prior active Army experience who must be retrained.[24] This problem is amplified in the Army Reserve because it has a much higher percentage of combat service support slots than does the active component.

Even soldiers within the same unit change specialties. Unlike the active component, there is no replacement "pipeline" in the reserve components from which to fill empty personnel slots. When the infantry company supply sergeant leaves the unit, the unit is not assigned another supply sergeant; instead, it often assigns one of its infantrymen to fill that position and retrains him. A veteran reserve soldier who has held two or more military occupational specialties is the rule rather than the exception. Unless the definition of unit stability is strictly the length of time the same soldiers have been members of the same unit, reserve component units may not be as stable as we are led to believe, even discounting turnover.

As a result of this inefficient system, as of fiscal year 1990, 10 percent of Army National Guard soldiers and 11 percent of Army Reserve soldiers were not qualified in their specialties, and an additional 8 percent of the Guard and 16 percent of the Reserve may have been qualified but were not working in the specialties for which they had been trained.[25] More than one-fifth of the soldiers in Selected Reserve units, therefore, are unqualified to perform the duties of the positions they hold. In 1989, the Reserve Forces Policy Board reported that individual skill qualification is the greatest limiting factor for the Army National Guard, and the second most critical for the Army Reserve.[26]

The promotion system in the reserve components also inhibits readiness. The reserve component officer promotion system differs from that of the active component. In the active component, officers are selected by Department of the Army promotion boards that consider all officers based on their date of rank (i.e., the date they were promoted to their current grade). Although this system is used in the reserve components as well, the reserve system has an additional qualifier: reserve component officers must satisfy a minimum time-in-service criterion. For example, to be considered for promotion to major by a Department of the Army selection board, a reserve captain must have at least seven years in grade as a captain and twelve years of commissioned service. Unlike the active component promotion system, the statutorily mandated reserve component system does not permit recognition of the cream of the reserve officer crop by promoting them before their contemporaries.

This system places officers with prior active Army service at a disadvantage. A competent active component captain who resigns her Regular Army commission and accepts a reserve commission may have ten years of active service and more time in grade than most other reserve captains, and in fact may have already been selected for promotion in the active component. That officer is perhaps more qualified in her specialty than many reserve officers of senior rank. Nonetheless, although her active component contemporaries will be considered for major shortly after the captain's separation from the active Army, the newly appointed reserve captain will be forced to wait until she has twelve years of commissioned service before she can be promoted in the reserve components. Not only does this statutory inequity penalize reserve officers with substantial prior active service, but, more important, it fails to utilize their experience to the maximum advantage of the reserve components.

In fairness, the Reserve Officer Personnel Management Act would alleviate these inequities by promoting officers based solely on their time in grade and by instituting "below the zone" promotions (i.e., promotions ahead of an officer's contemporaries) for outstanding officers. The act, however, has not been high on the congressional list of priorities, despite its introduction in each Congress since the 100th. (The House actually passed the bill in 1992, but the Senate took no action.)

The Reserve Officer Personnel Management Act, though, would not change a facet of the active component officer personnel management system that has, in effect, been transplanted into the reserve components. The "up-or-out" system mandates that an officer who is twice not selected for promotion must be separated from the Army. The primary reason a reserve officer is passed over is failure to meet military educational requirements.

The up-or-out system is well suited to the active Army and even to the Selected Reserve. However, it kicks true Individual Ready Reserve mobilization assets out of the Army. A captain who has not completed his officer advanced course (a requirement for promotion to major) is not qualified for promotion and indeed may not be major material. But he surely could fill a captain's slot on a stateside staff upon mobilization, even if he were forty years old, perhaps freeing a more qualified active component captain for a higher priority assignment. The up-or-out system is thus a bullet lodged squarely in the Army's foot, especially because it drives the train for Individual Ready Reserve training by making officer professional education, rather than mobilization-related training, the top priority.[27]

Another aspect of the reserve component promotion system has an even more significant effect on readiness. The unit vacancy promotion system in essence allows commanders, subject to certain irreducible minimum time in grade, to promote officers to fill specific slots within their units. This system is yet another product of the reduced manpower pools available to fill positions in reserve component units, a byproduct of geography. Although the unit vacancy system applies to both the Army National Guard and the Army Reserve, the Reserve tends to avail itself of the system less than the Guard, because of both the Army Reserve's closer formal ties to the active Army and the requirement that the Army Reserve Personnel Center, an operating agency of the chief, Army Reserve, approve a Reserve unit vacancy promotion based on the unavailability or nonselection of Individual Ready Reserve officers to fill the slot.

In the Guard, on the other hand, a state nominates an officer for a unit vacancy promotion. That officer is then considered by a federal recognition board, which determines if the officer is minimally qualified for promotion, based primarily upon whether the officer has met the military education criteria for advancement to the next grade. Federal recognition boards for unit vacancy promotion nominations are virtual rubber stamps.

The effect of this system is to increase greatly the power of the states over officer personnel management. The system allows Guard officials to surround themselves with the officers they choose. In contrast to the active component, which promotes officers based on their performance and potential as reflected in their service records, the unit vacancy promotion system allows the Guard to determine its own criteria for promoting officers. In the case of the Virginia Guard, in which I served, loyalty was—explicitly—the foremost criterion for unit vacancy promotions, taking priority over both demonstrated performance and potential for service in the next higher grade. Through this system, many loyal and dedicated officers are promoted ahead of more competent (and perhaps even more senior) officers who would be considered better qualified for the job by the active component. Readiness suffers because reserve component units are not always led by the most able and qualified leaders.

In short, the unit vacancy promotion system abets cronyism, and cronism affects readiness. Even though many reserve component commanders and officers take advantage of the system to promote competent officers whose contribution to increased combat readiness is indisputable, the potential for abuse is great. Many officers succumb to the temptation to "look out for number one" by surrounding themselves with yes men at the expense of the greater good of their units. The system is a prime contributor to intra-unit politics, because playing politics increases the chances of receiving a unit vacancy promotion.

Of course, many of the same forces are at work in the active component officer corps, but they are not nearly as pervasive as in the reserve components. An active officer can inflate an officer efficiency report, but the check in the active component system is that the officer's next efficiency report will be written by someone else. In a system wedded to geography, a reserve officer may be continually rated by the same officer, and, in any event, efficiency reports often are not considered when commanders nominate officers for promotions to fill unit vacancies.

Aside from the fact that the unit vacancy promotion system may overlook competent officers, geography also limits the contributions those officers can make. Often, there is a limit to the rank an officer can attain in a troop unit because there are no vacancies in the next higher grade in units close to that officer's home. Thus, to quote frequent reserve critic William V. Kennedy, "We are . . . training, at great expense, thousands of Reserve and National Guard officers each year, most of whom will be denied full military careers because the Army . . . National Guard and Reserve and their individual units are run as tight little geographic fiefdoms."[28]

Closely related to the promotion system is the fact that reserve component units recruit their own personnel. This system, especially for officers, also brings politics and cronyism to the fore. Officers are often accepted into units based not on their service records but on recommendations and civilian-style job interviews—on who they know rather than what they know. Unlike the active component, which assigns officers to fill vacant slots, the onus is on the reserve officer to find his own slot. This system creates a greater incentive to "network" than to be thoroughly competent. Like the unit vacancy promotion system, the reserve component unit recruiting system allows commanders to set their own personnel standards, rather than following the Army's.

One flaw in the reserve component personnel system is unique to the Guard: the Army National Guard commissions its own officers. The officers of the Army's other two components are commissioned through resident programs (with the exception of those officers, mostly doctors and other medical professionals, who receive direct commissions), either the United States Military Academy, ROTC, or the Branch Immaterial Officer Candidate Course (commonly called officer candidate school) at Fort Benning, Georgia.[29] The states, on the other hand, operate their own officer candidate schools on the same one-weekend-per-month, two-weeks-per-year basis as other Guard training. The states commission a significant number of their officers from their own officer candidate school programs; as of December 1991, close to one-third of the Guard's officers were state officer candidate school products.[30]

These state commissioning programs affect significantly the quality of the Army National Guard officer corps. Although the programs are run according to a standard Department of the Army program of instruction, there are inevitable disparities among the various state programs, and deviations in techniques, if not standards, between the state programs and the active Army program at Fort Benning are unavoidable. Moreover, in contrast to every other commissioning program, in which cadets and officer candidates interact frequently with active Army officers, the Guard officer candidate programs are run by the states, insulating them from the active Army. Finally, there is no escaping the fact that Guard officers commissioned through state officer candidate schools are products of a part-time program replete with all the training detractors inherent in the reserve component system. One must be truly shortsighted to contend in good faith, as many Army National Guard supporters do, that the Army National Guard commissioning programs produce officers of the same quality as the full-time commissioning programs.

The full-time support force also impairs readiness. The most obvious effect is the shortage of full-time support personnel. As of the end of fiscal year 1992, only 70 percent of required Army National Guard full-time support billets were filled, and only 71 percent of equivalent Army Reserve slots were filled.[31] This shortage is a result of an age-old Army dilemma: more requirements than funds.

There are, however, less apparent but equally negative effects inherent in the current full-time support system. Except for assigning active component personnel to reserve component units (and the number of active soldiers so assigned is minuscule), the current full-time support system exacerbates the distinctions between the components. Although AGR soldiers cost every bit as much as active component soldiers of the same grade and are fully expected to meet and maintain the same professional standards, beneath the surface, AGR soldiers are not comparable to their active Army counterparts.

In the Army National Guard, the states, in practice, manage AGR soldiers. There are a finite number of AGR slots within each state, and these slots require disparate specialties. Therefore, to fulfill the normal progression of assignments, AGR soldiers, like their part-time counterparts, must change specialties frequently. For example, an AGR field artillery captain, assigned to the G-3 section of a division staff, may find himself as the S-3 (operations officer) of an infantry battalion when he is promoted to major, in spite of the fact that he has no infantry training or experience. The S-3 slot may be the only available AGR major's slot in the state.

The Army Reserve AGR program suffers the same woes, even though it operates on a larger scale. Army Reserve AGR officers are managed by the Reserve's Full-Time Support Management Center in the same manner that active component officers are managed by the Total Army Personnel Command. They are thus rotated among Army Reserve units and headquarters throughout the country. The pool of Army Reserve AGR officers, however, is tiny in comparison to the active component officer pool. When the Full-Time Support Management Center reassigns an AGR officer to a vacant slot, the assignment creates a chain reaction; for every assignment, a new vacancy is created. This chain reaction often results in the assignment of at least one AGR officer to a position for which that officer is unqualified.

The professional competence of AGR officers is further hindered by the fact that they are often denied the opportunity to command units because command in the reserve components, as a rule, is reserved for part-timers. Soldiers who remain in the AGR program continuously may find themselves filling positions for which, in the active component, command is normally a prerequisite.[32] These soldiers, however, simply because of their full-time status, are denied the benefit of command experience.

Full-time support personnel are administrators and managers. Their units assemble only one weekend per month and an additional two weeks per year. Full-time reserve soldiers are thus desk-bound, keeping their units running between drills and planning training and other events. In contrast, active component personnel in comparable positions execute as well as administer and plan. In addition to their administrative and managerial capabilities, active officers and noncommissioned officers routinely employ their technical, tactical, and leadership skills.

Full-time support personnel, therefore, tend to be less competent technically and tactically than their active Army counterparts. Indeed, full-time support personnel—on whom the part-timers rely heavily for technical and tactical competence, simply by virtue of their full-time status—may be no more capable in those areas than the part-time reservists whom they see only weekly or less, even though AGR soldiers are theoretically the equivalents of active component soldiers. In fact, many part-timers with relevant prior active component experience may be considerably more skilled in their specialties than the full-time support soldiers who wear the uniform to work every day at the armory or Reserve center.

Unit readiness suffers because of this phenomenon. Because they wear the uniform to work every day, the remainder of the unit expects more of the full-time support personnel than they can deliver. Many unit commanders (nearly all of whom are part-timers) often rely to their detriment on less-than-satisfactory technical and tactical advice from their full-time support staff. The result is flawed plans and faulty execution, and, ultimately, degraded readiness.

Virtually all of the reserve component readiness detractors combine to affect the professional development of reserve leaders and produce what one officer has euphemistically called "a lack of expertise in command and staff areas."[33] This characterization significantly understates the problem; many senior reserve officers and noncommissioned officers are simply unqualified for the positions they hold. According to the General Accounting Office, noncommissioned officers in one Army National Guard unit it examined lacked initiative, discipline, and proficiency in basic soldier skills.[34]

The pillars of the Army's leadership development program are institutional training (formal military and civilian schools), operational assignments, and self-development.[35] There is simply less time for self-development in the reserve components; reserve soldiers work two jobs. But reserve component leaders are severely disadvantaged in the other two areas as well.

Although reserve component and active Army soldiers undergo the same basic combat training upon entering the Army, a part-time reservist obviously cannot attend subsequent resident active component military education courses—institutional training. These courses, therefore, are often "exported" to the reserves—that is, the reserves are provided a program of instruction and they run these courses locally. U.S. Army Reserve Forces schools, for example, offer reserve officers the opportunity to attend the Command and General Staff College in local communities on a part-time basis. Once the active component passes the programs of instruction to the reserve components, however, the active component washes its hands of reserve individual education.

In addition to exported training, the active Army operates resident reserve component courses at active component service schools; these courses are usually truncated versions of active component courses. This training enables reserve soldiers who can attend a six-week course, but not a six-month course, the opportunity to do so. Reserve component soldiers also can enroll in correspondence courses, many of which entail a two-week resident phase. For all of these reserve component courses (many of which are available to active Army soldiers as well), the reserve soldier receives the same educational credit as an active component soldier who attends the analogous resident active Army course. Full-time support personnel can, but often do not, attend full-length resident active component courses.

Except for, arguably, the truncated reserve component versions of active component resident courses, the reserve component individual military education program is inferior to the active component resident program. Instructors in U.S. Army Reserve Forces schools generally lack the breadth of expertise and experience of active component instructors, and that a correspondence course is an inadequate substitute for a six-month or one-year course is beyond cavil.

Reserve component leaders are also disadvantaged in terms of the development they receive through their experience in units—the operational assignment branch of the professional development process. Clearly, they get only thirty-nine days' worth of such experience per year. Perhaps more important, reserve soldiers do not receive the professional coaching and guidance they need from their superiors. This is a direct reflection on the quality of senior reserve component leaders. Senior leaders are the products of the reserve component political system as much as they are products of the military system. They may be dedicated and loyal, but they are not necessarily competent.

The military expertise of these senior reserve officers (lieutenant colonels and above) and noncommissioned officers (sergeants first class and above) clearly is hindered because they do not have the luxury of practicing their profession every day like their active Army counterparts. Many reserve component senior leaders have active component experience, but they gained that experience as junior officers or enlisted soldiers. Although professionally invaluable, very little of that junior-level active component experience translates directly to senior-level leadership. For example, in any active Army division G-3 section, there is a wealth of higher-level staff experience among both the officers and noncommissioned officers. In the Army National Guard division G-3 section in which I served, there was only one soldier who had served on an active component staff above battalion level, and none had ever served on a division staff before the 29th Infantry Division was activated.

Senior leadership is learned from senior leaders. The barriers between the components prevent reserve component leaders from regularly observing the leadership, command, and staff practices of senior active component leaders who practice their craft daily. The result is that senior reserve leaders often are left to fend for themselves, to rely on junior-level active Army experiences, or to rely on past experiences that may be based on the poor examples of reserve senior leaders who were themselves not competent. This result contrasts with the active component, where senior leaders can call on past experience gained from observing their own leaders, both good and bad, day in and day out in tactical units.

As but one example of a senior reserve component leader who had not learned senior leadership techniques, I observed a reserve component division commander who habitually slept at night away from his headquarters with his infantry companies. Although his attempt to "lead from the front" was commendable, an active Army division commander would not consider absenting himself from his headquarters for entire nights, even in a training environment, because such absence removes the commander from his communications and his staff, severely impairing his ability to make decisions and thus to command his division. The senior leader's actions in this case were more appropriate for a battalion commander than a division commander, although even a battalion commander would not routinely spend the night away from his headquarters.

The senior leadership problem is self-perpetuating: poor senior leaders beget more poor senior leaders. More significantly, the quality of the reserve component senior leadership permeates the chain of command. Because they lack the day-to-day tactical and military leadership experience of active component officers in comparable positions, many reserve senior leaders are unable to impart their experience informally to their subordinates, depriving other reserve component leaders of the solid experiences they should be gaining through operational assignments.

The difficulty of implementing the Army's leader development program within the constraints of the reserve components contributes to reserve leadership shortcomings. And, clearly, the reserve components' leadership shortcomings affect readiness—commanders who cannot prepare mission essential task lists and commanders who lack the experience to plan and conduct effective training.

Of necessity, the reserve component force is a reflection of the civilian population, and this fact in itself impedes the readiness of the force. Unlike the active Army, the civilian population is notoriously unfit physically, and this is reflected in the reserve components. Although the reserves are ostensibly held to the same fitness standards as active Army soldiers, physical fitness is just not a routine part of the civilian work day, as it is in the Regular Army. Although reserve component soldiers may be more fit than the civilian population as a whole, they are not as fit as active component soldiers. In combat units, this fact is especially significant. Moreover, reserve component leaders do not emphasize physical fitness to the same extent as their active counterparts because the reserve unit that assembles once a month has substantially less control over the physical fitness of its soldiers than the active Army unit that conducts physical fitness training regularly.

On the average, reserve component personnel tend to be older than active component personnel, and, although the Reserve Forces Policy Board has stated that this age differential is insignificant,[36] it does affect unit readiness in some instances. Although this tendency may be advantageous in some cases, as a rule it is not. In combat arms units, physical qualities are at a premium, and older soldiers simply cannot keep up. A tank commander depends on his reflexes to identify and hit targets. An infantry platoon sergeant needs the strength and stamina to lead his soldiers on forced marches carrying a full combat load on his back. Even more senior leaders must be capable of the physical vigor necessary to "lead by example." In some cases, these positions in reserve component units are filled by soldiers in their mid- to late forties. Some noncommissioned officers in the Georgia Army National Guard's 48th Infantry Brigade (Mechanized), mobilized for Desert Storm, were well over fifty; the brigade's command sergeant major was fifty-four, and another senior noncommissioned officer was fifty-eight.[37] These soldiers, although they are not too old to contribute to the Army, are too old to be counted on to serve effectively in leadership positions in combat units.

The final detriment caused by the reserve component system is also the least apparent, and its existence is frequently denied both by the active and the reserve components. The reserve components and the active Army are distinctly separate cultures. Active component soldiers who have not served in a readiness group or in some other assignment that brings them into daily contact with reserve soldiers (in other words, the majority of active component soldiers), tend, at least among themselves, to disparage the reserves as unprofessional. Active Army soldiers typically do not understand the constraints under which the reserve components operate. The typical active soldier "lack[s] a generic understanding of the history, purpose, and nature of the Reserve components with whom they must work, train, and fight alongside upon mobilization."[38] Colonel David E. Shaver of the U.S. Army War College sums up the situation by stating that "the [active component] officer seems to possess an *arrogance* toward, and an *ignorance* of, the [reserve components]."[39]

This phenomenon is present even at the highest levels of the Army, because, although senior active component leaders understand the reserves intellectually, rarely have they experienced the reserves' challenges first hand. As a senior reserve officer explained: "Few current senior Army leaders have had substantive personal service with [r]eserves. Due to disestablishment of the readiness regions, an [active component] major general command, there is today no seasoning assignment for [active component] brigadier or major generals. There are colonel assignments in readiness groups or as senior advisers, but promotions from those positions have been rare."[40]

This lack of understanding has made the reserve components, in the words of Congress, "second class citizens."[41] Although the phenomenon has diminished considerably since the advent of the Total Force Policy and will continue to do so in the wake of the mobilization of the reserve components for Desert Storm, it will exist as long as the components remain separate in practice, if not theory.

The cultural difference is significant, because it cements the barriers between the components and flies in the face of the Total Force Policy, which, after all, "integrates" the active and the reserve components into a cohesive whole. As Martin Binkin and William W. Kaufmann state in their book *U.S. Army Guard and Reserve: Rhetoric, Realities, Risks,* perhaps the seminal study to date on the reserves: "Deep feelings . . . have traditionally marked relations between the Army's active and reserve factions, a natural cleavage between professional and citizen soldiers that widens in the face of a shrinking availability of defense resources or a growing perception that reserve forces are substitutes for—rather than complements to—the active forces."[42]

All of the inefficiencies and inequities in the current system combine to produce reserve component combat units that the Army cannot readily provide to the combatant commands, and reserve component combat support and combat service support units that are mission capable but deficient in many respects. The Army must correct deficiencies in—or in some cases, wholly rework—these units after mobilization before they are ready to go to war.

Although the Total Force Policy envisions some postmobilization training, it requires units to achieve premobilization expectations. These expectations in fact are often not met. The ultimate result of the three-army system is that reserve component resources are wasted in peacetime by inefficiently preparing units that are not as combat ready as they could or should be, and active component resources are wasted after mobilization by correcting reserve deficiencies that should not be there in the first place.

These wasted resources—principally time and money—are precious and irreplaceable commodities that, in the foreseeable future, will remain in short supply. We cannot afford to waste them. Some day the cost may be more than money.

5. LESSONS LEARNED FROM DESERT STORM

> The most important military problem is to devise
> means of preparing great armies of citizen soldiers to
> meet the emergency of modern war.
> —John McAuley Palmer

OPERATION Desert Storm was the first—and only—true test of the Total Force Policy; never before have the reserve components been mobilized on such a scale and thrust into action so quickly. The reserves played a key role in Desert Storm, and the operation proved that some facets of the Total Force Policy worked well. Other facets, however, did not work at all. Because Desert Storm was both short and unique, we must not be lulled into a false sense of security by the generally successful performance of the reserve component support units and personnel who were deployed to Southwest Asia.

Although some conclusions concerning the reserves' performance in Desert Storm can be drawn, just as important are the conclusions that cannot. Problems remain, even though they may not have surfaced or may have been ignored during the war. Therefore even the reserves' successes in Desert Storm must be critically analyzed, for the next war may not be fought against a third-rate enemy.

Desert Storm began as Desert Shield in August 1990, when Iraq invaded Kuwait. In response to this aggression, the United States began to deploy active component forces from the United States to Saudi Arabia, under the command of the United States Central Command, the unified command with responsibility for the Southwest Asia theater.

On August 22, 1990, President Bush authorized the 200K call-up, and thus gave the services the authority to mobilize Selected Reserve units and personnel. Many reserve component units deployed to the Persian Gulf, while others served at Army bases in the United States and in Germany to keep those bases operating in the absence of deployed active component units.[1] In

January 1991, the president declared a national emergency, and the Army mobilized more than twenty thousand members of the Individual Ready Reserve.[2] In all, approximately 150,000 reservists from both the Army National Guard and the Army Reserve served on active duty during Desert Shield and Desert Storm.[3]

Significantly, the majority of the reserve component units deployed to the Persian Gulf were combat service support units, and, except for two Army National Guard field artillery brigades, the remainder were combat support units. Three Guard roundout brigades were activated but were not deployed to Southwest Asia.

The Desert Storm mobilization validated some aspects of the Total Force Policy. The "stay behind" rate for reserve units, that is, the percentage of a unit that did not deploy with the unit for one reason or another (usually medical, in the case of the reserve components), was roughly equal to the active component nondeployable rate.[4] Clearly the reserve soldier was ready and willing to serve.

Moreover, Desert Storm vindicated the Total Force Policy in hometown America. Although there were certainly factors other than the reserve component mobilization that contributed to the popular support of the Persian Gulf war, the fact that communities throughout the United States were affected by the mobilization of their reserve component units without doubt helped sustain the nation's patriotic fervor. As it was designed to do, the Total Force Policy ensured that the Vietnam experience did not recur on the home front.

Finally, Desert Storm proved, at least in the specific context of that conflict, that reserve component combat support and combat service support units can mobilize, deploy, and accomplish their battlefield missions. As General Colin L. Powell, chairman of the Joint Chiefs of Staff, stated, the reserve components "were vital to the success of the early stages of the operation. . . . It is safe to say that without them the swift and efficient deployment of our forces would not have been possible."[5] By almost all accounts, reserve combat support and combat service support units supported the force well.

Nonetheless, there were deficiencies, even in combat support and combat service support units. In general, the physical condition of reserve component soldiers was less than adequate; up to

80 percent of some mobilized California Army National Guard units, for example, were unable to meet Army physical fitness standards.[6] Moreover, combat support and combat service support soldiers often had been trained to perform their support missions at the expense of basic battlefield survival skills. Thus, when they were mobilized, mechanics were trained to fix vehicles and medics were trained in operating room techniques, but many of these soldiers were out of shape and could not put on their chemical protective masks or decontaminate themselves in the event of a chemical attack. As a result, many reserve component units consumed prime postmobilization training time accomplishing tasks that should have been accomplished before mobilization.[7]

The Army ignored these deficiencies in reserve component combat support and combat service support units and deployed them anyway, for several reasons. First, the units were essential to the accomplishment of Central Command's mission. The first reserve component units mobilized were those that provided capabilities that the active component does not possess. Deploying those units even with their shortcomings was better than lacking their capabilities entirely. Second, reserve component medical, transportation, water purification, and other support units are expected, for the most part, to operate in the corps rear areas and in the theater communications zone, away from the "front lines." The battlefield survival skills of these units, although not unimportant, are less critical than those of front-line combat units. This was especially true in Desert Storm, because Iraq's ability to strike these rear areas was virtually nullified by the total ineffectiveness of its air force. Finally, most support units tend to be more stationary than combat units, and the remainder rely almost completely on vehicles to move about the battlefield, placing less of a premium on the physical fitness of soldiers. In the end, the reserve combat service support and combat support units got the job done. They were not as prepared for war as active units, but in Desert Storm this fact proved to be less than critical.

Other incidents during the mobilization cast doubt on the efficacy of the Army's readiness reporting system as it applies to the reserve components. The readiness reporting system reports personnel readiness (e.g., assigned strength), equipment readi-

Table 4. Army Training Readiness Levels

Readiness level	Estimated days needed to be fully trained
C-1	0-14
C-2	15-28
C-3	29-42
C-4	More than 42

Source: GAO, *Army Training: Evaluations of Units' Proficiency Are Not Always Reliable* (Washington, D.C.: 1991).

ness (e.g., that on hand and operable), and training readiness. Readiness ratings range from C-1 (zero to fourteen days required to be fully ready) to C-4 (nondeployable; there is also a C-5 rating for units undergoing equipment transitions) (see table 4). Personnel and equipment readiness can be measured objectively; training, of necessity, is assessed subjectively by the unit commander.

During Desert Storm, the Army selected units for mobilization based in large part on their reported readiness. Yet many units that were ready on paper were in fact unable to perform their missions. Personnel were unqualified for the positions they filled, but were approved—in the case of some Army National Guard units by the *states*—to fill those positions. Because they were considered acceptable substitutes, these personnel (e.g., a dermatologist in a battlefield surgeon's slot) did not detract from their units' readiness ratings.

Interestingly, many, if not most, reserve component unit commanders downgraded their units' subjective ratings to C-4 upon mobilization. Although some of these changes may have been legitimate, as in the case of a unit that lost key personnel immediately before mobilization, the obvious inference is that the commanders inflated their original subjective assessments of the state of their units' training. Although this surely occurs in the active component as well, the effect is much greater in the reserve components because active component units are better trained than reserve units. Because a unit rated C-4 is not deployable, whereas units with ratings of C-3 through C-1 are, there is a big difference between a unit whose readiness rating is inflated from C-2 to C-1 and one whose rating is inflated from C-4 to C-3. Desert Storm proved Binkin and Kaufmann's assertion that "the training ratings have an optimistic bias, reflecting the reserve unit commander's 'can do' attitude as much as his unit's readiness."[8]

More significantly, Desert Storm marked the utter failure of two key reserve component programs: CAPSTONE and roundout. CAPSTONE was completely disregarded during the Desert Storm mobilization. The Army did not mobilize reserve component units according to their wartime CAPSTONE affiliations as outlined in Central Command operation plans, but on their reported state of readiness.[9] The major purpose of the CAPSTONE program, to align units during peacetime so that they will be prepared to fight together during war, was negated. Thus units that had trained with active component Central Command units were not mobilized, whereas units with CAPSTONE missions for theaters other than Southwest Asia were.

Third Army's mission is to serve as the Central Command Army component headquarters. Stationed at Fort McPherson, Georgia, the Third Army commander, a lieutenant general, wears a second hat as the deputy commander of Forces Command. When deployed as the Central Command Army component headquarters, Third Army commands all U.S. Army units assigned to Central Command. These forces include the various corps assigned to the command and "echelons-above-corps" units, or units not assigned to the subordinate corps and therefore controlled directly by Third Army.

Before Desert Storm, the major echelons-above-corps commands projected to be assigned to Third Army upon deployment to Southwest Asia were reserve component units. Third Army's theater army area command, its two-star rear area support command, was the 377th Theater Army Area Command, an Army Reserve headquarters. Third Army's two-star signal command was the 335th Signal Command, headquartered in Atlanta, within five miles of the Third Army headquarters at Fort McPherson. Both of these commands had planned and trained extensively for combat in the Middle East. Both commands had interacted frequently with Third Army and had deployed on the Bright Star series of exercises to the Southwest Asia theater.

Neither of these two headquarters, however, was mobilized. Instead, the Army formed a provisional theater army area command, piecing together an ad hoc headquarters from the Forces Command staff and other units. The Army also created a provisional signal command with personnel from the U.S. Army Information Systems Command, a nontactical command based in the United States. Undeniably, these provisional commands performed ably; the commander of the provisional theater army area command, Lieutenant General William G. Pagonis, in fact won wide acclaim and a third star.

It is certainly true that the failure to mobilize the 377th Theater Army Area Command, the 335th Signal Command, and other headquarters and units with Southwest Asia CAPSTONE affiliations is at least partially attributable to the provision of mobilization authority in increments that restricted the number of reservists that could be mobilized and to the uncertainty surrounding the date ground combat in the Persian Gulf would commence.[10] Nonetheless, the failure to mobilize these units provides unmistakable evidence of the uselessness of the CAPSTONE program. One reserve general officer rationalized the failure to mobilize these units by asserting that the Army has an obligation to exhaust active component resources before throwing civilian communities into disarray by mobilizing reserve component units.[11]

The Total Force Policy, however, supposedly ensures that the reserve components are no longer the "second string" but are to deploy and fight alongside the active component. CAPSTONE is the program that purports to implement the Total Force Policy. Although perhaps the Army envisioned CAPSTONE in the context of full mobilization, theoretically the program should apply to lesser mobilizations as well. Theory aside, creating provisional commands when there exist reserve headquarters that could do the job is not cost-effective; for example, the Army spends five million dollars annually to train and maintain the 335th Signal Command.[12]

Not only did the Army not mobilize the higher-level reserve headquarters aligned with Central Command through CAPSTONE, it generally did not mobilize senior reserve component headquarters at all, preferring instead to mobilize only those headquarters' subordinate units and to "plug them in" to deployed active component headquarters. Again, this fact evidences the Army's distrust of the CAPSTONE program. Many in the reserve component community conjured up a sinister active Army plot to avoid ordering senior reserve commanders to active duty in order to create more opportunities for senior active commanders to "punch their tickets." Although there may be some truth to this theory, the more plausible conclusion to be drawn from these events is that they provide unmistakable evidence of the active-reserve culture difference at work; justifiably or not, the active component lacked complete confidence in senior reserve commanders and believed that provisional active component units under active Army commanders would do a better job.

This conclusion evinces several explanations concerning CAPSTONE and the three-army system, all of which bring to the fore various problems enumerated in the previous chapters. First, Desert Storm underscored the separation between the active component and the reserve components. CAPSTONE did not work to bridge the gap between the three armies. Had it worked, active component gaining commanders would have known their CAPSTONE reserve component units well enough to have either insisted on their mobilization or found a more ready replacement unit well before mobilization.

Second, Desert Storm aptly demonstrated the active Army's lack of involvement in the CAPSTONE program. Had the system ensured that wartime gaining commanders took their peacetime CAPSTONE tasks seriously, reserve component units in general would have been more ready than they were, and CAPSTONE, rather than reported unit readiness, could have been the major factor in deciding which units to mobilize. One wonders if the deficiencies in individual battlefield survival skills and physical fitness, two areas that receive extensive emphasis in the active component, would have been as pronounced had the system fostered greater active-reserve interaction before mobilization rather than allowing the reserve components to circumvent the active Army guidance and advice they did receive.

The upshot of the combined failure of CAPSTONE and the readiness reporting system was that the Army did not know the quality of a reserve component unit when that unit rolled through the front gate of its mobilization station. Because the habitual active-reserve relationships envisioned under CAPSTONE never materialized, the Army depended at least initially on readiness reports in determining which reserve component units to mobilize, rather than on the wartime gaining commanders' personal knowledge of the reserve units' readiness. The systemic breakdowns resulted in mobilized units that looked relatively solid on paper, but in the flesh may not have been as ready as advertised.

The second systemic casualty of the Persian Gulf war was the roundout concept. Under that program, the Army assigned reserve component roundout combat units to active component divisions. The theory behind roundout was that those reserve units were to deploy with, or shortly after, their parent active divisions, and certainly were to fight with them.

Two of the divisions deployed to Southwest Asia had reserve component roundout brigades. The 1st Cavalry Division's roundout brigade was the 155th Armored Brigade of the Mississippi and Texas Army National Guards. The roundout brigade for the 24th Infantry Division (Mechanized) was the 48th Infantry Brigade (Mechanized) of the Georgia and South Carolina Army National Guards. Neither division deployed with its roundout brigade, nor did the brigades subsequently join their divisions in the Persian Gulf region. The 1st Cavalry Division deployed to Saudi Arabia with a brigade from the 2d Armored Division attached in lieu of the 155th; the 24th Infantry Division deployed with the 197th Infantry Brigade, an active component separate brigade, attached in lieu of the 48th. Interestingly, the 24th Infantry Division and its roundout brigade were equipped with the M-1 tank, but the 197th had the older M-60.

The Army did not mobilize the roundout brigades at the outset of the conflict and apparently intended not to mobilize them at all, at least in part because of the statutory constraints concerning the length of time they could remain on active duty. The Army's failure initially to mobilize the roundout brigades generated considerable controversy. The Department of Defense, according to General Powell's testimony before the Senate Armed Services Committee, maintained that mobilizing the roundout brigades was infeasible because they would require significant postmobilization training to be prepared for combat in the Persian Gulf.[13]

Congress obviously disagreed with the Department of Defense's decision. The fiscal year 1991 Defense Authorization Act stated in a "sense of Congress" provision that "the President should order to active Federal service at least one Army National Guard combat brigade for deployment in the Persian Gulf region in connection with Operation Desert Shield."[14] The political volatility of the issue was further exemplified by the National Guard's assertions that the Army's failure to initially mobilize and then deploy the roundout brigades was the Department of Defense's attempt to play politics: by not utilizing the Guard combat brigades, so the Guard argued, the Army could make its case for retaining active component force structure at the expense of the reserve components in the coming drawdown.

The Army finally mobilized the 48th Infantry Brigade on November 30 and the 155th Armored Brigade on December 7, four months after Iraq invaded Kuwait. In addition, the Army mobilized the 256th Infantry Brigade (Mechanized) of the Louisiana and Alabama Army National Guards, the roundout brigade for the 5th Infantry Division (Mechanized), which did not deploy to the Gulf. The 48th Brigade was sent to the National Training Center for postmobilization training.

The 48th's experience at the National Training Center brought the futility of roundout, heretofore untested, directly into the public spotlight. The brigade was kept at the National Training Center longer than any other unit had remained there previously and was still training there when the ground campaign of Operation Desert Storm ended. Although neither the Department of Defense nor the Army would officially acknowledge the fact, the brigade remained at the National Training Center because of significant training shortcomings. Their deficiencies when first mobilized included lack of adequate technical, tactical, and leadership competence among officers and noncommissioned officers at all levels; physically unfit personnel, in many instances because of age; and lack of qualification on equipment and weapons under field conditions.[15] Maintenance was also a serious problem; according to one report, the brigade could keep only 30 to 40 percent of its tanks operational at the National Training Center.[16] Undoubtedly, the brigade's maintenance problems were related directly to the fact that its soldiers could not maintain their vehicles at home because of the complexity of the M-1 tank; because the brigade's subordinate units are located significant distances from Fort Stewart, Georgia, where the vehicles are located; and because much of the brigade's peacetime maintenance is performed by state civilian employees.[17]

The Army did not certify the brigade as combat ready until three months after it was mobilized, including two months of intensive training at the National Training Center—and even after it was certified, only about half of its tank and fighting vehicle crews met gunnery qualification standards.[18] The unit had such severe problems that the brigade commander, a Georgia Army National Guard brigadier general, was replaced, ostensibly by the Georgia Army National Guard, after the brigade was ordered to active duty. Because Georgia Guard officials do not have the legal authority to relieve an Army National Guard officer in federal service, however, the Army must have played a role in the officer's relief, indicating the Army's opinion of how the brigade commander had prepared his unit for war. The plight of the 48th Brigade received widespread attention from the national news media.

In addition to the problems that befell the 48th Brigade, the 256th Infantry Brigade received unfavorable publicity on the network evening news as well. After mobilization, the Army sent the 256th, along with the rest of the 5th Infantry Division, to train at Fort Hood, Texas. While at Fort Hood, some sixty-seven soldiers from the brigade's 1st Battalion, 156th Armor went AWOL (absent without leave). Among the reasons the AWOL soldiers gave for their actions were low morale, inadequate training, and stressful conditions. Allegations of poor leadership were also made.[19] Whatever the reasons, the soldiers' actions indicate that the unit had problems. If the unit could not handle stressful conditions at Fort Hood, they surely were not prepared to cope with the rigors of combat. As further evidence of the brigade's unpreparedness, eight of its company commanders were replaced during training.[20]

By any standard, the roundout brigades were not prepared to deploy with their parent active component divisions. Moreover, they were still not prepared to deploy even a month after they were mobilized. Two conclusions can be drawn from these facts. Either the Army knew that the roundout brigades were not combat ready and did not expect them to go to war with their divisions, or the Army had faith in the roundout brigades' readiness and later discovered that they were not as ready as the Army had expected them to be. There is at least some evidence to support the view that the Army never expected the roundout brigades to be ready to deploy. Stephen M. Duncan, then the assistant secretary of defense (reserve affairs), implied in testimony before the Senate Appropriations Defense Subcommittee on April 9, 1991, that the Department of Defense had planned all along that roundout units would receive enough postmobilization training—more than thirty days—to make them combat ready, despite operation plans that relied on roundout brigades being in the combat theater thirty days after mobilization.[21]

If the Army did not expect the roundout brigades to be ready, the obvious inference is that roundout is a sham, a sop to the National Guard and the politicians who support it. If the Army's decision not to deploy the roundout brigades was simply the result of the brigades' failure to attain the expected level of premobilization readiness, one must deduce that there were problems with the Army's readiness reporting system as it pertains to the Army National Guard roundout brigades (or that the roundout commanders overestimated their readiness), or that the expectations for the roundout brigades were unrealistically high. In either case, whether roundout is a sham or whether the program was simply too ambitious and the brigades failed to meet the Army's expectations, the conclusion that the concept was flawed from its inception is inescapable.

After Desert Storm, the Army continued to cling to the party line that roundout is viable. General Edwin H. Burba, then commander in chief of Forces Command, testified before the House Armed Services Committee that the roundout concept can work and that Guard combat units can be ready for battle thirty days after mobilization. General Burba's assessment, however, reflects the dutiful soldier supporting the chain of command.

After General Burba's pronouncement, Secretary of Defense Dick Cheney changed the position of the chain of command, opining that roundout did not work in Desert Storm and, accordingly, that the concept should be reevaluated. Cheney said that the Defense Department's expectations of the Guard roundout brigades were unrealistic. Lieutenant General James W. Crysel, the Second Army commander, subsequently stated that a minimum of ninety days of postmobilization training is required before roundout brigades should be deployed.[22]

Moreover, even some in Congress began to reconsider their prior support of the concept. Representative Les Aspin (D-Wisconsin), then chairman of the House Armed Services Committee (and now secretary of defense), questioned the ability of roundout units to be ready to fight in time to get to the battle with their active component divisions. Of course, Secretary Aspin's views will not be shared unanimously, and the staunch Guard supporters in Congress—who invariably seem to share the same parochial view of national defense as the Guard itself—will disagree with him.

Indeed, roundout continues to receive support. General Robert Sennewald writes that the 48th Brigade was ready to go to war one hundred days after mobilization, and he deduces from this result that roundout remains a viable concept. General Sennewald asserts that roundout is advantageous because "it provides an opportunity for the [reserve component] elements to work closely with the active force and assimilate the latest Army doctrine, tactics and procedures."[23] General Sennewald neglects to emphasize, however, that the 48th had the resources of the entire National Training Center to itself for three months and that 4,370 active component soldiers were assigned to train the 48th Brigade (in all, nearly nine thousand active soldiers were committed to training the three mobilized roundout brigades).[24] If roundout continues, what happens when more than one roundout brigade needs that type of training? The resources simply won't be there. Moreover, if the Total Force Policy truly worked, *all* reserve component units, not just roundout brigades, would be able to interact closely with the active component to "assimilate the latest Army doctrine, tactics, and procedures."

The Department of Defense has charged that "the attention given to the readiness of the . . . roundout brigades was way out of proportion" because the three mobilized roundout brigades constituted only 7 percent of the total number of reserve component soldiers called up.[25] But only three roundout brigades were mobilized because the war against Iraq was a mismatch. True, under the current system, reserve component combat units are less critical in the short run than reserve combat support and combat service support units. Reserve combat units, however, are by no means unimportant. In Desert Storm, the active Army was an eighteen-division force with four highly trained divisions forward deployed in Europe; two of those divisions in Europe deployed to the Persian Gulf. When the active Army has reduced its rolls by six divisions and 250,000 soldiers (and the National Guard Association would cut two more divisions from the active force structure and transfer them to the Army National Guard), the Army's approach to future Desert Storms will be considerably different—it will need reserve component combat forces on the battlefield.

The debate over the performance of the roundout brigades in Desert Storm pointedly illustrates the political forces surrounding the reserve components, and especially the National Guard, at work. Secretary Cheney's pronouncement regarding the reevaluation of roundout set off a hailstorm of protest from the National Guard. South Carolina's adjutant general, for example, blithely stated that the 48th Brigade's tank battalion, a South Carolina Army National Guard unit, was combat ready, despite the overwhelming evidence to the contrary.[26] Moreover, the South Carolina adjutant general echoed the earlier sentiments of the Guard establishment, accusing the secretary of attempting to save active component force structure by impugning the readiness of the Guard.[27] Another Guard official defended the roundout program by saying, "If you're going to relegate [the roundout units] to poorer cousins, if they're not going to have proper equipment, then why should anybody join [the Guard]? Why not just draft me whenever you get the urge?"[28] The official publication of the National Guard Association, *National Guard,* rationalized that readiness criteria changed after mobilization; Guard roundout units, the magazine argued, were ready by the premobilization standards.[29]

The statements of these Guard officials and supporters aptly demonstrate the tunnel vision and typify the shortsighted views of the National Guard discussed earlier. Legitimately concerned about self-preservation, they view any proposal to diminish the Guard's role not in its larger—and intended—context as a potential method of enhancing the national defense, but as a threat to the Guard's existence. Desert Storm brought out the worst of a politically dominated reserve component system that only secondarily responds to the imperatives of national defense. Deploying combat units that were less than adequately prepared would have been the height of irresponsibility. Once again the Guard proved that many of its members and supporters are more concerned with defending the Guard's performance than with improving it.

The largest lesson from Desert Storm is that a significant difference exists between the capabilities of reserve component combat support and combat service support units and those of reserve combat units. Reserve component combat support and combat service support units, despite some shortcomings, were thrust into the fray and did the job. With the exception of the two Army National Guard field artillery brigades that were deployed, the same cannot be said for reserve component combat units.

This difference was completely predictable.[30] Support units, by their very nature, depend to a much greater extent on the individual capabilities of their soldiers than do combat units. In this regard, they are much like the Air Force's reserve component units that performed so admirably during the Persian Gulf war.

A combat service support unit is a collection of individuals who work largely as individuals. Although certainly combat service support units require a modicum of collective training to be proficient, especially at the higher staff level, complex coordination and teamwork of the type required of maneuver units is not necessary for a mechanic to fix a truck or for a driver to drive one, or for a doctor to treat a patient. Moreover, combat service support tasks are easily transferable to the civilian sector. In fact, many combat service support soldiers get *more* experience performing their military duties in the civilian work place than they do in their military roles; civilian doctors in urban trauma centers, for instance, have more experience treating gunshot wounds than most active-duty surgeons. Although large-scale maneuver exercises are helpful in training combat service support units, they are not essential; most essential combat service support training can be conducted in the armory, the motor pool, or an accessible local training area.

In contrast, infantry and armor units (and field artillery and other combat units, to a lesser extent) are, in General Sennewald's words, "faced with a special training readiness challenge—the complex mission of synchronizing the AirLand Battle and associated training problems."[31] These units must conduct maneuver exercises to be even minimally proficient. Combat units rely extensively on coordination and teamwork; comparing the training required of an infantry unit to that required of a medical unit is a little like comparing the type of practice necessary to meld a winning football team versus that required to create a successful track team. Not only must infantry and armor soldiers be proficient in their individual military occupational specialties, they must be able to operate as part of an infantry squad or a tank crew. Moreover, the squads and platoons and companies must be able to operate as part of battalions and brigades and divisions. The type of training necessary to produce effective combat maneuver units requires both the time and terrain that are generally unavailable to reserve component units.

The Desert Storm mobilization also left many reserve component questions unanswered. Although some Individual Ready Reserve soldiers were mobilized and performed well, they were called up in insufficient numbers to test the Individual Ready Reserve system adequately. Because only a relatively small number of Individual Ready Reserve personnel were required, the Army was justifiably selective in choosing the soldiers it called up. Therefore, aside from medical personnel, whose military competence is directly related to civilian job skills, the Army mobilized the Individual Ready Reserve soldiers who had most recently separated from active duty and who thus required little postmobilization training. Additionally, most Individual Ready Reservists the Army called up possessed critical combat service support specialties and were assigned to combat service support units, and the mobilization of the Individual Ready Reserve as a system of casualty replacement in combat units or of initially bringing combat units up to their authorized strength was not exercised. The conclusion that the Individual Ready Reserve system worked during Desert Storm and that previously identified problems have been solved is therefore unwarranted.

Desert Storm was the Total Force Policy's maiden flight. Undoubtedly, the policy will be refined and, hopefully, improved. In any event, Desert Storm was a microcosm of virtually everything that is both right and wrong with our current system and provides an excellent springboard from which to mold a citizen-soldier force for the twenty-first century.

When considering the readiness of the reserve components, Desert Storm must be put in proper perspective. The overall success of the reserve component combat support and combat service support effort must be qualified and viewed in the context within which it occurred; even support units had problems, but these problems were masked by the brevity and uniqueness of Desert Storm. If the Army can be guaranteed an opponent like Iraq in every future conflict, then the successes of the reserve components in Desert Storm can be accepted and their shortcomings rationalized away, and reform is unnecessary. If, however, the reserves are expected to prepare for tougher battlefields and more formidable opponents, then the Army must view the performance of even the successful reserve component units critically, and must make changes that minimize the chances that Desert Storm deficiencies will recur.

6. REALIZING THE POTENTIAL OF THE RESERVE COMPONENTS

> Our militia law is obsolete and worthless.
> —Theodore Roosevelt

HISTORY and politics maintain a stranglehold on the Army's reserve components, impairing their ability to function effectively as integral parts of the Total Army. Notwithstanding the rhetoric of the Total Force Policy, the active and reserve components do not mesh; rather, according to a former chairman of the Joint Chiefs of Staff, there exists a wall between the active component and the reserve components.[1] This wall inhibits the readiness of the reserves, a situation that, according to Binkin and Kaufmann, "would be less worrisome if many reserve units were not being counted on as equivalents of active-duty units, to be deployed early in a war."[2]

Despite the increased importance placed on the reserves by the Total Force Policy, the basic framework of the reserve components has changed little in forty years. The last significant modification of that framework was the Armed Forces Reserve Act of 1952. In 1952, though, "weapons and tactics were simpler, the reserves were not expected to be deployed early in war and certainly were not considered to be a leading force or to provide a timely response in crisis management, and much of the reserves' manpower consisted of World War II veterans."[3] Moreover, despite the significant changes in the art of warfare, Selected Reserve training requirements remain essentially the same today as they were forty years ago.[4]

The need for substantive change in the reserve components was summed up by former Department of Defense manpower and reserve affairs official James L. Lacy:

> There is little doubt that, were we able to begin afresh, even the most hallucinatory of military plans would scarcely come close to

producing the patchwork of reserve components we field today. Burdened with mysterious redundancies (Why, for instance, do we require both an Army Reserve and an Army National Guard . . . ?), heavy with undifferentiated structure, employing across-the-board training-time requirements better suited to 1939 than 1989, and expected to function as an expeditionary force according to an unprecedented (and to many commentators, wholly implausible) deployment schedule—the reserves have become the single greatest choke-point in U.S. military preparedness.[5]

Although Desert Storm may have proved Lacy's ultimate conclusion, at least with regard to combat support and combat service support units, to be somewhat of an overstatement, his premises are valid. Wholesale assessment and reform of the organization, missions, and training of the Army reserve components are indeed necessary. To quote Lacy again:

The list of areas of useful inquiry and potential benefit in reserve reform is enormous in scope and diversity, in part because the nation's reserves have escaped critical reappraisal for most of the last fifty years, and in part because the nature of the military tasks ahead of them are more demanding than at any other time in their history. . . . If, however, one were to list the most compelling national security issues warranting urgent, sustained, and comprehensive treatment by the administration and the Congress . . . , reserve reform should be near the top of the priorities.[6]

The stated aim of the Total Force Policy is to integrate the Army into a cohesive whole. Four-star pronouncements notwithstanding, the Total Force Policy has not accomplished this aim. Three distinct components remain. As William V. Kennedy has noted, the time has come "to ask why the taxpayer should go on paying for three separate administrative and logistics systems for soldiers . . . who wear the same uniform, ride around in the same trucks, tanks, and airplanes, and are paid from the same U.S. treasury."[7] The system fairly cries out for change—significant, substantive, perhaps even revolutionary change, not "fine tuning."

The three-army system can be modified so that the gap between the reserves and the active Army is narrowed and so that the deficiencies bared by Desert Storm will not haunt the Army when the reserves are next mobilized. But the gap can be narrowed only by integrating the active and reserve components in fact as well as rhetorically. Truly to merge these three separate components into one army requires the elimination of the differences among the components, or their minimization to the extent possible. Ideally, there should be no difference in the way the Department of the Army interacts with and controls the active component and the way it interacts with and controls the reserves.

Completely integrating the active and the reserve components is, of course, impossible. Any plan to modify current structure, systems, and policies must begin with the premise that the active and reserve components are inherently dissimilar and that many of their dissimilarities cannot (and should not) be eliminated. The fact that the reserves are a citizen-soldier force and the active component is a full-time professional army will always remain. Nonetheless, eliminating as many dissimilarities as possible and reducing the effect of those remaining is the only way to merge the three armies and thereby to increase the readiness of the reserve components. Senator Daniel K. Inouye (D-Hawaii) understated the problem somewhat when he declared that "the blending of active and reserve forces needs further refining."[8]

Even discounting state control of the Army National Guard, there is no other reserve force in the world like the U.S. Army's reserve components. Switzerland, for example, maintains no standing army at all, relying solely on a citizen-soldier force. The Swiss, however, require military training of all male citizens. Israel, which boasts probably the best reserve force in the world, also has a universal military training requirement, and all of its reservists have prior active service. Moreover, Israel and Switzerland are not confronted with the geographic complexity the U.S. Army reserve system must face.

The Army reserve components are dissimilar even from the reserve components of the other U.S. armed services. The Air Force's reserve components, the U.S. Air Force Reserve and the Air National Guard, typically have enjoyed high readiness rates. Congress has suggested that the structure of the Air Force Reserve, under which the chief of Air Force Reserve commands all Air Force Reserve forces, has been a key factor in achieving those high levels of readiness, and has implied that adopting a similar structure will have a like effect on the Army Reserve.[9]

But comparing the Army reserve components to the air reserve components is like comparing apples to oranges. The Army and the Air Force are fundamentally different. The Air Force relies heavily on machines—its aircraft. The Army, on the other hand, is an organization whose success depends on the interaction of people. As Binkin and Kaufmann have stated:

> Basing expectations for the Army reserve components on the successes achieved by the Air Force Guard and Reserve overlooks some important differences between the services. Although the Air Force units' need to operate and maintain sophisticated equipment would appear to be demanding, paradoxically Air Force reserve units have found it easier than, say, infantry units have to maintain their proficiency. More Air Force reserve units are collocated with active units; the logistics, maintenance, and administration support is an obvious advantage. The mobility of Air Force flying units allows them to use distant training facilities, so more realistic combat training can be carried out on weekend drills. The nature of the Air Force missions permits a greater concentration on individual training and proficiency, as opposed to the larger maneuver exercises necessary to simulate land combat activity.[10]

The fundamental differences between the Army reserve components and the air reserve components that make any attempted comparison of the services' reserves inapt is further illustrated by the fact that Naval Reserve carrier air wings also have maintained consistently high levels of readiness, despite the problems that have beset the rest of the Naval Reserve.[11]

Air reserve component pilots, obviously key players in the Air Force readiness equation, in many cases are more skilled than their active Air Force counterparts. The majority of those pilots have previous active Air Force service. Further, as reservists, pilots generally are not required to rotate between flying and nonflying duties; they tend to remain in flying units. Thus, air reserve component pilots frequently have more accumulated flying hours than active Air Force pilots. Moreover, unlike Army reserve component soldiers whose equipment is located an average of 130 miles from their home station, the Air Force Reserve and Air National Guard pilots can get into their aircraft on weekends and fly them.

There is no single model on which to pattern reforms of the Army reserve components. To be effective, therefore, any change to our tradition-bound and politically complex reserve system must be truly innovative. Evolution has reached its zenith; the current systems, policies, and programs have given all that they possibly can to the national defense. What is needed now is the implementation of revolutionary ideas. To quote former Department of Defense reserve affairs official James L. Gould: "As is true of all large organizations, the Guard and Reserve will achieve their full national defense potential only if opportunities for innovations are periodically and systematically examined. This presupposes an environment in which even 'unthinkable thoughts' are addressed so that innovative proposals, which are at once conceptually and practically sound, can be developed for review by senior policy makers."[12] Just as fine tuning a broken television set will not produce a clear picture, simply refining the current inefficient system will not make the reserve components maximally effective.

The most radical, most politically controversial, and most necessary step in reforming the system is to eliminate state control of the Army National Guard. Ours is the only country in the world that cedes direct and legal command of nearly one-third of its army to the nation's political subdivisions.

Peacetime state control over the National Guard is a vestige of the colonial militia system, which was designed for purposes far different from those for which the Guard is used today. The military of the framers' era was a defensive force, which existed only to defend the territorial United States against foreign invaders. The framers simply did not contemplate the use of the militia beyond the borders of the United States. Furthermore, the framers of the Constitution developed the militia system out of a deep-seated fear of standing professional armies; in fact, many of the framers conceived of the militia as the nation's primary defense force.

As a military force, the Army National Guard today bears little resemblance to the old militia. As an integral part of the United States Army, the Army National Guard is certainly more than strictly a defensive force. It is part of the armed forces of a global military power with significant offensive capability and, as Desert Storm demonstrated, the very real potential for employment outside the territorial United States. Unlike two hundred years ago, as a world superpower the United States can ill afford to rely on a citizen-soldier force as its first line of military preparedness.

Although the Guard's federal mission today bears virtually no resemblance to the role of the colonial militia, the infrastructure within which the Guard must perform that mission is basically the same as that of the militia. Peacetime state control of the National Guard remains. This system is the legacy of a citizen-soldier force designed for a different time, a different place, and a different mission. It is wholly unsuited to the United States Army of the twenty-first century.

The increased funding with which the Guard has been blessed by the Total Force Policy cannot alone buy the Guard's readiness. Notwithstanding the money, peacetime state control of a significant portion of the Army and well over half of its reserve forces impairs the ability of the Army to prepare those forces for combat. State control of the Guard guarantees that training for combat during peacetime will not always be the number one priority, and that when it is, training will not always be accomplished to the standards of the active component, which, in the end, bears the ultimate responsibility for employing the Guard in combat.

Although the Guard assuredly does have a legitimate state mission, that mission is considerably less important than its federal mission as a reserve component of the Army. In financial terms alone, only 10 percent of the Guard's resources comes from the states.[13] The remaining 90 percent comes from the federal government, to be spent preparing the Guard for its mission as part of the federal armed forces. Moreover, the Guard's state mission—disaster relief and law enforcement, primarily—does not require mechanized infantry brigades, M-1 tanks, Apache helicopters, or sophisticated electronic warfare equipment. In the words of the Gray Board, "To pose a threat of artillery, armor or air power against citizens of this country is hardly in keeping with any foreseeable situation, and not consistent with sound public policy."[14]

This is not the forum for a constitutional analysis of this proposal. Suffice it to say that permanent federalization of the Guard would not violate the Constitution because Congress's constitutional power "to raise and support armies" is exceptionally broad. Furthermore, although the Constitution leaves much of the responsibility for governing the militia to the states, it certainly does not require the federal government to pay for the militia. Federalization would not divest the states of a state right guaranteed by the Constitution; it would merely recognize that the framers fully expected the states to pay for their militias.[15] Were the Guard federalized, the states (given the appropriate legislative mandate) would be more than free to constitute militia forces, which could be governed "according to the discipline prescribed by Congress."

Streamlining the reserves by permanently federalizing the Army National Guard would entail the elimination of that component as we now know it. The Army National Guard should be merged with the Army Reserve. (Whether the remaining reserve component is called the Army National Guard or the Army Reserve is irrelevant. The Army National Guard, of course, is the senior of the two current components.) Such a merger would simplify the structure of the Army; the Army would consist of two components rather than three. The states should fund their own defense forces.

The benefits of adopting this solution would be substantial. Most obvious, and most important, the United States Army would directly command the Army National Guard forces currently commanded by the states. Shifting of command would allow the Army to direct its reserve combat forces, instead of advising them, and would invest the Army with a mechanism for enforcing its guidance that is more effective than simply cutting off federal funds. Eliminating fifty-four separate links in the chain of command would also promote better and more efficient communication between the Guard and the Army. The National Guard Bureau would become unnecessary, and the Army could communicate directly with its subordinate reserve headquarters, like it does with its active component commands, rather than channeling its communications through a superfluous staff agency.

Abolishing state control of the Guard would also enable Army National Guard commanders actually to command their own units. That a division or brigade commander charged with employing his unit in combat has no authority over his subordinate commanders during peacetime is ludicrous; that those units are "commanded" by committee is militarily perverse. Permanent federalization of the Army National Guard would purge this anomalous system. A merger of the two reserve components would also eliminate the unproductive rivalry between the Army Reserve and the Army National Guard.

Federalization also would solve many personnel problems. State commissioning programs would die with state control. The states would not control AGR personnel. The role of politics that is inherent in a system in which the top military officers are state political appointees would be lessened considerably.

Elimination of the Guard as a state-controlled force would save money as well. Federalization would facilitate the elimination of the Guard's complex federal support structure, such as the National Guard Bureau (assuming, of course, that were this plan to be implemented, it would apply to the Air National Guard also), the Army National Guard Personnel Center, the U.S. property and fiscal officers, and the state area commands. Cutting the states out of the loop also would enhance efficiency, giving the taxpayers more for their money. Federally paid Guard personnel, for instance, would not have to spend their time responding to state requirements and training for state missions. In sum, federalization of the Army National Guard would significantly strengthen the Total Army.

As explained earlier, many of the same flaws that hamper the Guard in its quest to satisfy the Army's readiness standards are present in the Army Reserve also, even though that component is a fully federal force. Federalizing the Army National Guard is only three-quarters of a solution. The structure of the Army Reserve also must be modified to better enable reserve units to achieve their readiness potential.

Currently, the Army Reserve units that will (and did) deploy in wartime are controlled in peacetime by administrative, not warfighting, headquarters. Those headquarters, USARC and the ARCOMS, exist solely to command and control Army Reserve units (this is also a major mission, although not the sole mission, of the CONUSAS, to the extent that USARC has not yet displaced them). Their challenge is enormous: to train a myriad of different types of units for employment in different theaters of operations by different wartime commanders. Like National Guard state headquarters, these Army Reserve headquarters will not employ on the battlefield the units they train, even though they bear the ultimate responsibility for training them for war.

Of course, at some level, there must be a nonwarfighting headquarters that commands and controls subordinate warfighting units. That level, however, should be as high as possible. The reserve system should foster the greatest possible interaction between, and integration of, units that will fight together.

That is precisely how the system operates for active component forces. A forward-deployed active component headquarters, such as Seventh Army in Europe, will be reinforced by Forces Command units in the early stages of a European conflict. Those units, although commanded by Forces Command, are attuned in peacetime to the requirements of Seventh Army, the Army component headquarters for U.S. European Command, a unified command. III Corps, for example, headquartered at Fort Hood, Texas, trains for war in Europe based on the guidance of the U.S. Army Europe commander in chief (who wears a second hat as the commander of Seventh Army) and stays in close contact with U.S. Army Europe. Although placing the corps under the command of a headquarters on the other side of the Atlantic Ocean is obviously infeasible in peacetime, the active Army system allows the corps to train to the guidance of its wartime command while under the peacetime control of Forces Command. Moreover, every subordinate unit of the corps trains to the guidance of the corps commander, who of course has the undisputed power to enforce that training guidance.

The same cannot be said of the current Army Reserve system. Through CAPSTONE, Army Reserve units ostensibly train to implement the guidance of their wartime gaining commanders. Assuming that CAPSTONE works as it is supposed to, that guidance is enforced by the ARCOM or USARC commander. Thus, unlike every active two-star commander, the ARCOM commander is in the position of enforcing training guidance he does not issue, and, because he may not appreciate the nuances of the wartime missions of all of the commanders who will gain the ARCOM's subordinate units, may not fully understand.

As with any argument, there is a response to this one. One ARCOM commander contended that he knows his subordinate units better than any other commander and thus is in the best position to train them. He alleged that Army Reserve units, if trained to the appropriate Army Training and Evaluation Program standards (the Army Training and Evaluation Program

THE POTENTIAL OF THE RESERVES 113

contains every combat task a particular type of unit could conceivably be expected to perform), should be able to perform their missions for any wartime gaining commander.[16] This argument is fatally flawed. First, it admits that CAPSTONE is broken. More important, it implies that training guidance from wartime gaining commanders, such as that which CAPSTONE was designed to provide, is unimportant.

This implication could not be more erroneous. "Battle focus" is an essential part of Army training doctrine. As the Army's principal manual on training states:

> Battle focus is an approach used to derive peacetime training requirements from wartime missions. Battle focus is a process to guide the planning, execution, and assessment of each organization's training program to ensure they train as they are going to fight. Battle focus is critical throughout the entire training process and is used by commanders to allocate resources for training based on wartime mission requirements. The implementation of this concept enables commanders at all levels and their staffs to structure a training program which copes with non-mission related requirements while focusing on mission essential training activities. Battle focus is a recognition that a unit cannot attain proficiency to standard on every task whether due to time or other resource constraints. However, a successful training program is achievable by consciously narrowing the focus to a reduced number of vital tasks that are essential to mission accomplishment.[17]

Because reserve component units have less time and fewer resources, battle focus is even more important to them than to active component units. Battle focus narrows the Army Training and Evaluation Program to the mission essential task list. A reserve component commander cannot narrow his training focus adequately if he trains to Army Training and Evaluation Program standards without receiving appropriate guidance from his wartime gaining commander. Training without adequate guidance from the wartime commander implies one of two possibilities: either the reserve commander will try to drink from the fire hose and accomplish all of his Army Training and Evaluation Program tasks, or he will assume a wartime mission. In either case, the reserve component unit will not be trained adequately, unless, of course, the reserve commander correctly ascertains the precise mission for which his unit is ultimately employed, an unlikely possibility given the fact that in Desert Storm the Army did not match reserve component units to previously identified wartime gaining commands. Indeed, the ARCOM commander's reasoning contravenes even the Army's own recommendation, which is that the reserves should "do less better."[18]

Because CAPSTONE has proved unworkable, a system that matches units to missions must be developed. Moreover, that system must provide an incentive to produce reserve component units whose readiness is as good in fact as it is on paper. The new system must also take into account the basic differences between the active and the reserve components.

Such a system can be implemented only by forcing wartime gaining commanders to become directly involved with the pre-mobilization readiness of the reserve component units they will gain. The creation of USARC will not solve the problem. Indeed, simply inserting yet another administrative, as opposed to warfighting, layer of command between Army Reserve units and their wartime commanders, as Congress has mandated, may even exacerbate current problems.

Wartime gaining commanders, therefore, should control "their" Army Reserve units absolutely with respect to training, the cornerstone of readiness. Training funds should flow through wartime channels, not through peacetime administrative channels. Readiness reports should flow through the same wartime channels. As a corollary, wartime commanders should write the efficiency reports of their subordinate Reserve commanders. Thus, with regard to training, wartime gaining commanders would perform the functions currently performed by several disparate agencies: the ARCOMS, the CONUSAS, the readiness groups, and the active component units associated with reserve component units through CAPSTONE.

Administrative headquarters, however, are necessary in the Army Reserve because of the inevitable administrative and logistical headaches that geography causes. USARC, moreover, is essential as a powerful Army Reserve spokesman at higher levels; the command is necessary to address the "second-class citizen" problem. USARC, therefore, through the ARCOMS, should command Army Reserve units, but only for administrative and logistical purposes. It should have no role in training. And, although the CONUSAS play a significant role in the mobilization system and therefore could not be eliminated completely, their role in the training system could (and should) be transferred to wartime gaining commanders.

Like federalization of the Army National Guard, this structure would streamline the active-reserve interface. The commander of a unit slated for attachment to Third Army would report directly to the Third Army commander on training matters in peacetime, rather than through an ARCOM and USARC. The reserves' peacetime training structure would mirror their wartime command structure.

This system would eliminate several current shortcomings. First, it would force frequent contact between Army Reserve units and their wartime gaining commands. The habitual relationships thus developed would facilitate the rapid integration of Reserve units into their superior commands on mobilization. Because wartime gaining commanders would be familiar with their Reserve units and their capabilities, these relationships would help ensure that Army Reserve units would actually be employed in wartime by the commands that were identified in peacetime as the Reserve units' wartime gaining commands, and that planned and trained for that eventuality.

Second, Desert Storm proved that the readiness of reserve component units is not always reported accurately, especially with respect to training, and that mobilizing the reserves based largely on reported unit readiness does not guarantee that mobilized units are actually ready. An integrated active-reserve training structure would give wartime gaining commanders an incentive to make certain that their Reserve units were well trained. Wartime commanders also would know the readiness of those units first hand. By giving wartime commanders the formal *responsibility* to train the units that will be subordinate to those commanders during war, and the authority commensurate with that responsibility, the system would ensure that the active Army takes its peacetime role vis-à-vis the reserves seriously, and does not treat that role as just another support "tasking."

This streamlined system would work well for Third Army and other gaining commands headquartered in the United States, but some adjustments would be necessary for units slated for assignment or attachment directly to overseas commands. Nonetheless, like the proposed structure for reserve component units

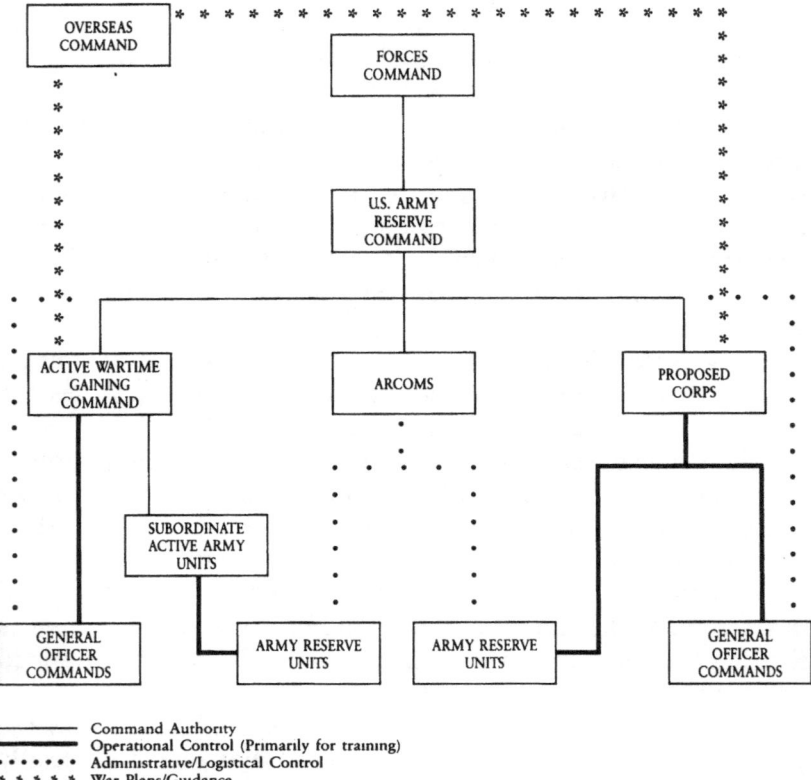

Figure 5. Proposed Organization for U.S. Army Reserve or Fully Federal Reserve Force

whose gaining commands are headquartered in the United States, the structure and relationships for reserve component units whose gaining commands are forward deployed should mirror the structure of the active component and should provide the same readiness incentives.

To this end, the Army should create new active component headquarters, with liberal reserve representation, and should give them the specific mission of training reserve component units (see fig. 5). These headquarters (call them corps, for lack of a better term) would be the stateside representatives of overseas commands and, with respect to training, would control all reserve component units slated for deployment to those commands. Their sole mission would be to ensure that their subordinate reserve component units are trained. Like active component Forces Command units with overseas wartime missions, such as III Corps, these corps headquarters would be directly subordinate to Forces Command and would coordinate regularly with the forward-deployed headquarters, serving as the conduits for training guidance based on the wartime commanders' concepts for employing their reserve units. These corps headquarters would issue and enforce their own training guidance to their subordinate reserve units.

The size and composition of these corps would depend on several factors. An overseas command the size of U.S. Army Europe may have two or three of these headquarters, and in that case, the corps could be organized geographically. A smaller overseas command, such as Eighth Army in Korea, perhaps would have only one corps.

To facilitate the interaction that would be required of scattered active and reserve headquarters, the headquarters would exchange liaison sections. At the gaining active component headquarters, the reserve liaison section would ensure that the wartime gaining commander and his staff fully appreciate the capabilities and the limitations of the reserve component unit. At the subordinate reserve component headquarters, the active component liaison section would communicate the wartime commander's intent and would serve as that commander's eyes and ears, compensating for the fact that the gaining commander may be across the country rather than across the street. Reserve component units slated for assignment or attachment to forward-deployed commands would exchange liaison officers with their newly created corps headquarters. This system, incidentally, would work equally well when a reserve component unit is the gaining command for an active Army unit.

The involvement of wartime commanders in the preparation of the reserve components for war is essential to achieving maximum reserve readiness. A new system that puts this burden squarely on those wartime commanders is the only way for the Army to train as it will fight. Reforming the active-reserve interface to ensure that the Army trains as it will fight—integrating the Army practically as well as rhetorically—will significantly enhance the Army's ability to provide trained reserve component units to the combatant commands.

The current Individual Ready Reserve policies and programs also bear reexamination. Even today, the Individual Ready Reserve is inadequate for its mission. But the Army of the twenty-first century will be considerably smaller than it is now, and this will magnify the importance of the Individual Ready Reserve.

Currently, only a relatively small percentage of the Individual Ready Reserve is sufficiently trained to be capable of serving immediately as individual replacements, especially in combat units where such personnel will be essential. But because the size of the Individual Ready Reserve is directly proportional to the size of the active component, as the size of the Regular Army shrinks so will the Individual Ready Reserve eventually contract, reducing the number of trained individual replacements even further. Although there are no publicized plans to do so, the Army could use Selected Reserve units to provide individual soldiers to fill vacancies in deploying units if absolutely necessary, as it has done in the past (indeed, the Army did this during the Desert Storm mobilization under the guise of "cross-leveling," or shifting personnel between units to bring them up to strength). As the Army shrinks, however, this option will disappear, because Selected Reserve units themselves will be mobilized earlier (assuming that combat and support units are distributed between the active and reserve components in the same proportion as they are currently).

The Army must increase its emphasis on the Individual Ready Reserve. Ideally, all Individual Ready Reserve soldiers should be required to train annually; at a minimum, those whose civilian occupations are incompatible with their military specialties require training. A required two-week training period for Individual Ready Reserve soldiers is not unreasonable. The Army could easily incorporate such a commitment into enlistment contracts without affecting active component recruiting because as the Army shrinks, competition for a reduced number of active component personnel spaces will increase. Moreover, this training period should be just that: structured, mobilization-related individual training rather than professional development schooling or the assignment of Individual Ready Reserve soldiers to active Army units in which they may or may not receive the training they need, depending on unit training schedules and the whims of the unit chain of command.

The up-or-out promotion system should be eliminated for the Individual Ready Reserve. Elimination of the up-or-out system for Individual Ready Reserve soldiers would allow the Army to shift the emphasis of Individual Ready Reserve training from promotion-oriented professional development schooling to mobilization-oriented training, and would retain mobilization assets who may not meet promotion criteria.

If an Individual Ready Reserve annual training period is fiscally impracticable, as the Army maintains,[19] the status of Individual Ready Reserve soldiers' training should at least be monitored. The Individual Ready Reserve screening program allows the Army to muster Individual Ready Reserve soldiers for one day each year. This one-day muster, however, is wasted. The Army can make much better use of this one day by requiring Individual Ready Reserve soldiers to report to a centralized location near their homes, rather than to the recruiting station on Main Street, and detailing active component personnel to conduct military occupational specialty proficiency tests for these Individual Ready Reserve soldiers. At a minimum, such tests would give the Army a considerably better picture of its mobilization posture than it has currently.

Finally, Congress must change the mobilization authority to reflect the importance of the Individual Ready Reserve, especially if budgetary constraints dictate that the Individual Ready Reserve remain a low peacetime priority. The Army must be able to call up the Individual Ready Reserve at the earliest stages of mobilization to begin the training that will be necessary to make Individual Ready Reservists deployable when the war starts.

A streamlined reserve system addresses unit training problems, but reserve reform must also address professional development. The limits of individual military education are a fact of life in the reserve components, an unavoidable consequence of a citizen-soldier force. Nonetheless, the quality and availability of individual reserve component military education can be improved.

Currently, to receive the best possible military educational opportunities, the reservist must avail himself of the active component educational system, but because of competing interests, most reserve component soldiers are precluded from doing so. Although the Army exports training from the active Army "schoolhouse" to the reserve components, this training is notably inferior to that received in the active component schoolhouse itself. The primary reason for this is that the exported training is conducted by the reserve components, or in some cases by active component Forces Command units, not by the U.S. Army Training and Doctrine Command, the Army subject-matter experts. For example, the Command and General Staff Officer Course instructors in U.S. Army Reserve Forces schools are not selected based upon their past experience or upon any particular expertise. They join the staff of the Army Reserve Forces school because the school happens to be a conveniently located Army Reserve unit.

Exported training should be conducted by Training and Doctrine Command, the same major command that conducts schoolhouse training for the active component.[20] There is no reason (other than money) that the Command and General Staff College at Fort Leavenworth, Kansas, cannot send instructors across the country to teach reserve officers. Similarly, the U.S. Army Armor School, rather than an active component Forces Command unit, should be responsible for conducting new equipment training for reserve component tank units converting from M-60 to M-1 tanks. Putting the onus on Training and Doctrine Command to *train* the reserve component soldiers instead of simply prescribing the courses of instruction would eliminate the Hobson's choice that now requires reserve soldiers to opt between attending active Army courses or receiving substandard instruction. An exported training system run by the subject-matter experts would ensure that schoolhouse training for the reserve components equates more nearly to that of the active component.

To integrate the active and reserve components effectively, personnel programs must be integrated also. Without addressing personnel programs, merging the two reserve components will not narrow the gap between the active and the reserve components, but simply will magnify the disparity between the current reserve system and that of the active Army. Although the issues

THE POTENTIAL OF THE RESERVES 121

addressed below involve officers, they apply, albeit perhaps to a lesser extent, to enlisted soldiers as well.

Current personnel programs in the reserve components are vastly different from those in the active component. An active Army signal battalion commander, for example, will have had repeated assignments in signal units to qualify him for his command, and indeed such assignments are prerequisites for that command. A reserve component signal battalion commander, on the other hand, is not required to fulfill the same prerequisites. On the contrary, the reserve component battalion commander may have spent his formative years as an armor officer.

Active component battalion and brigade commanders are selected for command based on their official records by Department of the Army command selection boards. The reserve component command selection process is decentralized. Reserve component commanders are most often selected by their superior commanders (although in the Army Reserve, the Army Reserve Personnel Center can have some input). This system elevates personal relationships—read politics—to a level at least equal to, if not greater than, professional competence.

Promotion standards in the reserve components are lower than those in the active component. Most reserve component selection boards select officers for promotion on a "fully qualified" basis (i.e., every officer who meets the minimum prerequisites is promoted), and there is no "below the zone" selection process to promote outstanding officers.[21] This system rewards mediocrity and fails to recognize outstanding performance. Finally, each component maintains its own personnel programs and manages its personnel through its own personnel agencies: the Total Army Personnel Command for the active component, the Army Reserve Personnel Center for the Army Reserve, and the Army National Guard Personnel Center for the Army National Guard (the Army National Guard Personnel Center is more of a records repository than a personnel management agency; the states actually manage Army National Guard personnel).

These differences between the components should be eliminated. Transfers between branches by reserve component officers should be carefully scrutinized, as they are in the active component, and the qualifications of reserve component officers for command and promotion should be similar to those of active

Army officers. Battalion and brigade commanders should be selected by centralized Department of the Army boards.

The unit vacancy promotion system should be abolished and a "best qualified" system of promotion should be adopted, even if fewer officers are promoted than are actually required. The reserve promotion system should recognize that some reserve component officers who do not meet the current minimum time-in-grade and time-in-service requirements are more qualified for promotion than many of the officers who do. By refusing to promote mediocre officers, the reserve components could fill key positions with the most qualified officers available, even if those positions called for officers of senior rank, rather than bypassing more capable officers of lesser rank because another officer met the bare minimum standards for promotion.

These objectives can be best accomplished by merging the components' disparate personnel management agencies. All Army soldiers, regardless of component, should be managed by a single centralized personnel agency to ensure commonality of programs, policies, standards, and administration. Consolidation of these disparate agencies would, of course, save money. Such a consolidation has been proposed, and, predictably, the reserves have opposed it.

Implementation of these personnel reforms would raise the quality of reserve component leadership significantly. And the quality of reserve leadership is directly related to the state of reserve readiness.

Reform of reserve personnel issues must also address the full-time support system. The current full-time support program suffers from many of the same ills as other reserve personnel programs: full-time soldiers without the requisite experience for the positions to which they are assigned and full-time support personnel with no greater tactical and technical competence than their part-time counterparts.

The military technician program is a logical one. It saves the government money by not allowing technicians access to the full panoply of military benefits. The technician program implicitly realizes that a reserve component administrator is not equivalent to a full-time active component soldier.

The AGR program, on the other hand, is not as logical. AGR

soldiers, for all intents and purposes, are identical to active Army soldiers in every respect but one: their technical and tactical competence. The Senate Armed Services Committee described the situation: "The compensation and benefit program available to full-time [R]eserve and [G]uard members is virtually identical to that available to active duty personnel, despite the different conditions of service between the two groups."[22] AGR soldiers are desk-bound; they have no more field experience than many of the part-time soldiers who look to the AGR personnel for technical and tactical expertise. In essence, the major difference between an AGR soldier and a technician is cost (although, to be sure, the Army has greater control over AGR soldiers because technicians are civil service employees), and the Army may get more for its money through the technician program.

Nonetheless, the AGR program is potentially a good one. Completely replacing AGR soldiers with active component soldiers will not work. An active component soldier will never fully comprehend the challenges facing the reserve components because the active soldier has never worked a civilian job during the week and attempted to get an Army unit ready for war on weekends. The AGR soldier understands these challenges.

Ideally, there should to be no difference between an AGR soldier and an active component soldier of the same branch and grade. The gap between the capabilities of active component soldiers and AGR soldiers can be spanned by routinely filling half of the positions currently occupied by AGR personnel with active component soldiers. This would infuse reserve component units with active Army tactical and technical expertise. Moreover, such a program would ensure a continual rotation of fresh ideas into reserve component units, eliminating the stagnation caused by the "that's-the-way-we've-always-done-it" attitude that often exists in the reserves because key personnel have risen to their positions from within the same unit.[23] As an additional benefit, assigning active soldiers to reserve component units would placate Congress before it legislates this solution if the Army fails to take the hint in the Senate's version of the 1991 Defense Authorization Bill, which, although the provision was not enacted into law, directed the Army to study the feasibility of such action.[24]

The AGR personnel displaced by these active component personnel should be assigned in the active component. Rotating AGR soldiers between the reserve components and active Army units would provide the reserve components a much larger pool of positions to fill with their AGR soldiers, allowing those soldiers to be assigned to positions best suited to their experience, qualifications, and professional development needs. Such assignments would provide a means to get AGR soldiers away from their desks and out into the field to train regularly with tactical units. Active component assignments would create entirely new possibilities for AGR personnel, such as the opportunity to command. When these AGR soldiers return to the reserve components, they too would infuse their units with fresh tactical and technical experience, and would be able to provide their units with the skills and expertise the part-time soldiers expect of them. Moreover, routinely assigning reserve component soldiers to the active component and vice versa would do much to eliminate the active-reserve cultural differences that exist due to infrequent contact between the components.

The first response to these proposed improvements will be that geography makes them impossible to implement. Indeed, personnel issues in the reserve components are influenced heavily by geography. In an infantry battalion headquartered in a relatively small community, the next battalion commander in all probability will be one of the two majors in the battalion, because there are only two infantry majors in the geographical area. Under these conditions, centralized command selection makes little sense. Along the same lines, reform of the reserve promotion system to make it more similar to the active component system would accomplish little in geographically isolated units, because there is an extremely small manpower pool from which to fill the unit's positions.

Geography, of course, affects other key aspects of reserve readiness. Distance from suitable training areas and dispersion of reserve component units inhibit the reserve components' ability to achieve their readiness potential. Thus, loosening the hold of geography on the reserves would remove many stumbling blocks on the reserve components' path to increased readiness.

Many reserve component combat support and combat service support units can achieve their readiness potential in spite of the geographic factors that affect other reserve component units.

A judge advocate general detachment or a combat support hospital, for instance, is affected little by geography. Normally, the reserve components can find enough reserve lawyers or doctors in the same area to man those types of units, because those soldiers perform the same tasks in their civilian occupations as they do in the reserves. Furthermore, those units rely little on large collective training areas and therefore can accomplish significant training on weekends.

For combat units, however (and perhaps for some combat support units, such as certain military police units), the geographic hurdle is one to be reckoned with. To improve their readiness, these units must have soldiers who are qualified in their military occupational specialties, and they must have easy access to suitable maneuver training areas and ranges. Moreover, the geography problem is closely related to the time problem, because distance from training areas coupled with the training time of only two days per month means that these combat and selected combat support units cannot conduct meaningful training on weekends.

To overcome the combined effect of the geography and time factors, combat and certain combat support units should draw their personnel from the entire national reserve manpower pool. Thus, a tank crewman separating from the active component who desires to continue his service in the reserve components would be assigned to an armor unit based largely on the needs of the Army, precisely the primary assignment criterion used in the active component. The soldier would not be forced to join his hometown reserve unit, which, if it is a transportation or maintenance unit, has no slots for tankers. Although geography would be a factor in assigning soldiers to units, it would not be the primary consideration.

This system would solve most of the problems inherent in the current geography-based reserve system. Soldiers would be assigned to units based on their qualifications, not on their place of residence, facilitating centralized command selection and a revamped promotion system. The frequent branch switching that presently occurs could be eliminated. Such a system would complement centralized personnel management and would lessen the effect of politics in the reserve components.

Of course, a system free of the current geographic constraints would not work without modifying the weekend drill system as well. Clearly a reserve component unit composed of soldiers from across the country cannot be expected to gather one weekend every month. For these units, the weekend drill system should be scrapped and replaced with a second two-week training period. To compensate for the decreased time that would be available for the part-time soldiers to devote to their reserve units, the number of full-time support personnel in these units must be increased to pick up the slack.

Abolition of the weekend drill system alone, even without nationalizing the manpower pool for combat units, would be beneficial. In William Kennedy's words, "Common sense alone dictates that the high degree of teamwork required of armor, infantry and artillery units and the larger and more complex brigades and divisions cannot be fully achieved on weekends and during [one] two-week training period, much of which is taken up with administrative matters."[25] The abolition of weekend drills for units that can accomplish little meaningful training during that time would allow those units to use their available training time much more profitably.

The elimination of weekend drills and the institution of two semiannual training periods of two weeks each would actually reduce the number of days reserve component units assemble. Nonetheless, because the best training is conducted at annual training, those units would accomplish much more in twenty-eight or thirty days at a good collective training facility than they can accomplish now during their two-week annual training period and an additional twenty-four days at the armory or Reserve center.

Units that are converted to the national manpower pool/two-by-two week semiannual training concept should be relocated to the installations at which they would conduct their semiannual training, or to the immediately surrounding civilian communities. This would accomplish two purposes: first, it would eliminate dispersion between units; second, it would save the two training days that are currently wasted deploying units to and from their annual training sites.

The implementation of such a system, of course, would have an impact beyond the Army. There is almost always tension between civilian and part-time military careers. Many employers already resent the amount of time their reservists devote to the Army, often at the expense of their civilian occupations. The demands of reserve service sometimes force reserve component soldiers to choose between their civilian careers and the military[26]—indeed, according to a 1988 Army Reserve study, spouse and employer conflicts are among the most important reasons that soldiers leave the Army Reserve.[27] A system in which civilian employers would be forced to cede their employees to the Army for twice the amount of time as is required currently perhaps would not draw widespread support among corporate America, and corporate America has a strong voice on Capitol Hill. Certainly a six-week total annual training requirement, a proposal floated by the Department of Defense,[28] would ask too much of civilian employers, who would be loath to hire an employee whose absence from the job for two months annually (six weeks of annual training plus two weeks of vacation) would be assured.

Nevertheless, its military advantages aside, the system is far from infeasible. In enacting parental leave legislation, Congress has mandated unpaid leave for periods longer than the four weeks that the semiannual training periods would require, and that legislation potentially affects a segment of the population much larger than the part-time reserve force. Moreover, because reserve soldiers would not drill during the month under this system and because many of them would live far from their unit headquarters (and thus could not "drop in" once a week), they would have more uninterrupted time to devote to their civilian jobs. And although willing and enthusiastic employer support cannot be legislated, Congress could provide financial incentives to employers whose reserve soldiers would train four weeks annually.

The enhanced readiness this system would produce must be balanced against the value of maintaining the presence of reserve units in American communities (although individual reserve soldiers would continue to show the flag in their communities even if their units were not there). A prime advantage of the hometown reserve system is the support that it generates among the American public. Therefore, this system should be implemented only for those units that are affected significantly by geography and time, primarily combat units. The system for other units would not change—the costs of uprooting field hospitals and water purification units from local communities outweigh the advantages.

This facet of the proposed system raises questions concerning the feasibility of implementing the various centralized personnel reforms proposed earlier for units that would continue to drill on weekends. Although geography obviously must be considered in implementing centralized personnel functions, the best and the brightest reservists are dedicated enough to commute an hour or two more, if not even longer, to their units (especially if they are compensated for their travel). As in the active component, promotion and command are powerful incentives in the reserve components; indeed, lack of advancement opportunities is a common reason for attrition.[29]

The proposed national manpower system raises yet another question. Extra income is a strong reserve recruiting incentive. Reserve component soldiers receive one day's pay for each four-hour unit training assembly. Thus, for a typical two-day weekend drill, reserve component soldiers receive four days' pay. During annual training, though, they receive one day's pay for one day's work, just like active component soldiers. Under the two-by-two week solution, therefore, many reserve soldiers would be paid for only twenty-eight or thirty days of duty annually rather than for sixty-two. These soldiers would receive no extra income (generally, civilian employers, at most, pay reserve soldiers the difference between their military and civilian pay during annual training periods).

In implementing separate systems for combat and support units, the compensation disparity between combat soldiers, who would be paid one day's pay for one day's work, and combat support and combat service support soldiers, who would continue to receive double pay for weekend drills, must be addressed. The answer is to offer combat arms soldiers enlistment and reenlistment bonuses, as in the active component, or to offer larger bonuses to those who would receive them under the existing system. These soldiers would get their extra income in a lump sum, as opposed to once per month. Such bonuses would also be a strong incentive for soldiers separating from the active component to continue their service in the reserve components.

The proposed system envisions new costs—enlistment and reenlistment bonuses, increased full-time support personnel in some units, funding to monitor Individual Ready Reserve training, travel for Training and Doctrine Command instructors, and financial incentives to employers. These costs beg the obvious question: How do we pay for them?

Obviously, detailed financial analysis is necessary to answer this question. Nonetheless, on an intuitive level, the system has inherent savings to offset the new costs. Eliminating or streamlining redundant organizations—state headquarters (through federalization of the Army National Guard), overlapping active and reserve component agencies (such as the Army Reserve Personnel Center and the Army National Guard Personnel Center), CONUSAs and readiness groups (by giving wartime gaining commanders responsibility for training)—would save substantial sums. Reducing the number of personnel in administrative headquarters by abolishing those headquarters' training functions, and transferring responsibility for reserve component institutional training to the active component (thus eliminating or streamlining, e.g., Army Reserve Forces schools) would also result in savings. The new costs could well be incurred without significant budgetary impact (and perhaps even with a reduction in required personnel end strength).

Even assuming that the reserve components can reach their potential—that is, if the system can be improved so that the reserves can perform at their maximum capability—that potential must be defined. In other words, are some missions simply beyond the capabilities of a citizen-soldier force, and if so, what missions can such a force reasonably be expected to accomplish?

Under the current system, "there is precious little time between mobilization and deployment for the [reserves] to train, organize and establish command and control."[30] Desert Storm illustrated that this short amount of postmobilization time is simply inadequate to allow reserve component infantry and armor units to deploy with and fight alongside their active component counterparts in the early stages of a conflict. The Persian Gulf experience pointedly demonstrated the limitations inherent in the current reserve system.

Even in an ideal system, however, those reserve component maneuver combat units still would not have been able to contribute. The Total Force Policy expects too much of the reserve force, and consequently the reserves cannot help but fall short of expectations. As Utah's adjutant general has written, "Unfortunately . . . , there seems to be a philosophy of 'let's have all the capability or none,' rather than having a capability that is limited in deliberate ways to ensure compatibility with [actual] Guard and Reserve capabilities."[31] Reserve missions must be constrained by practicalities, and not assigned to reserve component units by virtue of the mere fact of their existence.

Although the Army apparently is moving toward consolidating contingency missions in the active component, the Total Force Policy has envisioned much of the reserve components as part of a contingency force, to mobilize rapidly and deploy almost immediately thereafter, and to fight with the active Army. As Desert Storm demonstrated, reserve component combat support and combat service support units may be up to this task, although improvements in their readiness are still necessary. Retention of an extensive combat support and combat service support capability in the reserve components is wise—such a force mix guarantees that the active Army will not ignore the reserves. Indeed, General Creighton Abrams, the Army chief of staff at the inception of the Total Force Policy, envisioned a Total Army in which the active Army could not support itself on the battlefield, so that a conflict fought without the reserves—another Vietnam—would be impossible.[32]

The contingency mission, however, exceeds the real capabilities of reserve component combat units. Reserve component infantry and armor units cannot possibly be fully trained before mobilization, or even thirty days thereafter.

Unit training is a building block process, and individual soldier skills are prerequisites for conducting effective collective training. Unit training starts at the lowest level, the infantry squad or tank crew, and continues progressively through higher levels. To conduct an effective company-level field training exercise requires trained squads and platoons, which in turn require trained soldiers.

But training is a perishable commodity, and statistics prove that individual reserve component soldiers are not well trained. Because its training time is both limited and infrequent, a reserve unit must devote a considerable portion of available training time to developing and maintaining individual proficiency. "Available training time," however, does not equal all the time a unit is together—thirty-nine days per year in the present system. Much of that time is occupied by the inevitable "training detractors": administration, inspections, HIV testing, and family support activities, to name but a few. All of these activities are necessary and enhance readiness in their own ways, but their negative effect on training is unarguable.

Reserve component combat units, therefore, must conduct a substantial amount of postmobilization training to enable them to deploy and fight effectively. Assuming that a reserve unit is as well trained as it can be before mobilization, and further assuming that the unit has qualified and competent leadership, even thirty days of postmobilization time would be inadequate, given that perhaps, at most, three weeks of those thirty days could be devoted to training. Time would be required to mobilize, travel to the mobilization station, prepare to train, maintain equipment, and prepare for deployment. Given the best of circumstances under current policies, a reserve combat unit thirty days after mobilization would have had perhaps five weeks of unit training during the previous year. To deploy a combat unit to war under those circumstances could be disastrous.

Thus, with the exception of vital combat support and combat service support units, the reserves should have a reinforcement mission, to deploy as follow-on forces to the active Army.[33] To expect reserve component combat units to be the equals of active component units is a goal too lofty for the reserves to reach, even with thirty days of postmobilization training.

The concept of roundout brigades, therefore, should be scrapped. Although reserve component combat service support units should retain their contingency missions to maintain the vitality of the Total Force Policy, reserve component combat units, in general, should be second-echelon forces. Reserve component infantry and armor divisions and brigades should be employed as reinforcing, follow-on forces, behind active component combat units.

In the wake of Desert Storm, the Army has realized the limitations of reserve combat units by restructuring the roundout program; two active divisions that before the Persian Gulf conflict were rounded out by Army National Guard brigades have each added a third active brigade. Nonetheless, even limiting the assignment of roundout brigades to active divisions that have reinforcing missions themselves is a questionable proposition. The resources necessary to train more than one roundout brigade simultaneously and the fact that the necessary postmobilization training would detract from the brigades' ability to plan for deployment and employment as part of their parent divisions—and indeed, the parent divisions' active component resources will have to be diverted to train their roundout brigades—may prove the roundout concept unworkable even for active component reinforcing divisions. These reasons may even doom roundout on the battalion level (although the Marine Corps successfully integrated company- and a few battalion-sized combat units into its active forces during Desert Storm).[34]

But eliminating roundout and assigning reinforcing missions to the reserves must not minimize their role in the Total Army. Second-echelon combat forces are important; any conflict larger than Desert Storm will require them, and, with a twelve-division active force, even a smaller conflict may require them. The reserve components, therefore, need not, and should not, be second-class forces.

The reserves must be trained to the standard the Army expects them to reach before mobilization so that they can meet realistic postmobilization timetables. Because of the difficulty of attaining and maintaining individual and small unit readiness, larger-scale reserve component premobilization training is usually unproductive; as a retired general officer has pointed out, "During annual training . . . , centralized training—often single echelon, small-unit training and evaluation—is the most effective, efficient training strategy."[35] Therefore, reserve combat forces should focus their premobilization training at lower levels, perhaps company and below, with the goal of reaching a level that will enable them to begin higher-level training immediately upon mobilization. Premobilization training should be oriented toward attaining an acceptable—and realistically achievable—level of training at the time of mobilization, and postmobilization training should allow the unit to coalesce into an effective fighting force. At a minimum, reserve component soldiers must be qualified and proficient in their military occupational specialties before mobilization. (The Army just recently has adopted this long overdue reform; the chief of staff directed that Army National Guard training should focus on soldier, crew, squad, and platoon skills.[36] Before Desert Storm bared the reserves' weaknesses, the chief of the National Guard Bureau publicly derided such a proposal.)[37]

In addition to second-echelon reinforcing forces, the Army must possess the capability to constitute third-echelon forces rapidly in the event of full or total mobilization. The cadre division concept, proposed by Congress and others,[38] and recently adopted by the Army, fits this mission ideally. Cadre divisions are to consist of skeleton staffs that could accept and rapidly train individual soldiers, thus forming combat-ready divisions that would deploy after second-echelon reinforcing reserve units. Under the Army's plan, cadre divisions will be skeleton Army National Guard units to be formed by thinning the ranks of existing Guard divisions.

There are, however, other possibilities. Units that could perhaps be more readily converted to cadre status currently exist: the Army Reserve training divisions. A reorganization of the training divisions and a reorientation of their focus could accomplish the transition to cadre divisions easily. The present mission of these stateside units is to conduct initial entry training—to train soldiers as individual replacements for deployed or deploying units. In the event of a conflict requiring less than full or total mobilization, the cadre training divisions would retain their current mission of training individual replacements. For full or total mobilization, they would train soldiers as members of their own deployable units.[39] Should full or total mobilization occur, the divisions' stateside training mission could be shifted to mobilized retirees.

Converting training divisions into cadre units and retaining their current mission would forestall the fear that cadre divisions, as they are currently described, would suffer from the morale problems that accompany the idleness and lack of resources inherent in low-priority units. Reserve component cadre divisions would continue to conduct basic and advanced individual training during peacetime, and the addition of active component soldiers would allow these units to expand their focus to encompass the tactical employment of the divisions during wartime.

Implementation of these innovations will require a redefinition of the Total Force Policy. The current concept envisions the active Army and the reserves as equal partners in all respects. But the components are inherently unequal. Only time and other resources can bring reserve component combat units to the active component's level of readiness, and time, at least, exists as a finite commodity for the reserves. No amount of money or legislation can create more time. This is not to say, however, that resources for the reserve components should be reduced or that the reserves should be de-emphasized; on the contrary, to improve the reserves so that they can meet more realistic expectations requires more resources, not fewer.

The Department of Defense's Total Force Policy Report, issued in 1990, is a step in the right direction, recommending that the active Army be capable of fighting conflicts of thirty days or less without reserve component support. Reserve combat forces, the report states, should not be used for contingencies lasting less than sixty days.[40] The Army evidently concurs with the recommendations of the Total Force Policy Report, because its new strategy calls for the active component to be capable of performing contingency missions of up to thirty days without any reserve component support.

But the Army may have taken a step too far, because relying completely on active component forces for contingency missions risks relegating the reserves to second-class status and losing the gains in reserve readiness we have made since 1973. The all-active contingency force subverts the basic premise of the Total Force Policy. Even assuming planners are omniscient enough to discern that a contingency will last for only thirty days, critical reserve units, even if they do not deploy, must be mobilized early on not only to guarantee that they are ready when and if they are needed, but also to assure public support for the operation.

In Desert Storm, public support for the war did not truly coalesce until the reserve components were mobilized. As Colonel Charles E. Heller of the U.S. Army War College has written: "Revert to a primarily [active component] contingency force for about 30 to 60 days without substantial [reserve component] augmentation and you run the risk that the 'will of the people' might never solidify around what some might perceive as the administration's 'military adventurism' using a 'mercenary' (All-Volunteer) Regular force. Involving the [reserves] early in any operation, including a contingency, is absolutely essential for the well-being of the Army and its fragile relationship with 'the people.'"[41]

Aside from the public support issue, the hard fact is that if the Army plans to fight contingency operations with only active forces, it may have to fight those operations without the important capabilities that are found only, or almost completely, in the reserve components—enemy prisoner of war military police units, graves registration units, and civil affairs units, to name a few (of course, some of these functions could be transferred in part to the active component, but this would have to be done at the expense of active combat units). Many capabilities, after all, were placed in the reserve components precisely because the architects of the Total Force Policy wanted to ensure that wars—even short wars—would not be fought without the reserves.

Obviously, all the differences between the components cannot be eliminated, nor should they be. The road to maximum reserve readiness, however, is to eliminate as many of the distinctions between the components as possible. This can be done only through reforms that create a reserve structure that mirrors the active component in all possible respects while maintaining the necessary distinctions between a regular, professional force and a part-time citizen-soldier force. Although such reforms may be painful and difficult to implement, they are necessary if the Army is truly committed to making the Total Force Policy work. And the Total Force Policy must work if the Army is to succeed on potential twenty-first century battlefields.

7. THROUGH THE POLITICAL MINEFIELD

> The slightest whiff of change sets in motion the well-organized, hometown political organizations of the competing Reserve and state National Guard systems.
> —William V. Kennedy

HISTORY and politics are the roots of virtually everything that is wrong with our reserve system. They are also formidable obstacles in the path of true reform. In the words of a former White House staffer, "These citizen soldiers are so solidly entrenched politically that no one in Washington dares challenge them frontally."[1] When we had a 780,000-member active Army, such a challenge was unnecessary. Cutting that force to 535,000 sounds a call to arms.

Historically, the reserves have vehemently opposed any change that would diminish their status within the military community. The National Guard in particular has adamantly stood on its soapbox to prevent any diminution of either of its identities, as a state organization or as a key federal reserve force. The Guard has traditionally had its cake and eaten it, too: it has remained a state institution while becoming the Army's primary reserve combat force.

Guard and Reserve soldiers are unabashedly congressional constituents, and they are not averse to letting Congress know it. Not surprisingly, Congress has traditionally lent a sympathetic ear to the reserve lobby, the National Guard Association and the Reserve Officers Association. So, starting from the premise that the Guard in particular will oppose substantive reform if the object of that reform is to assign a more realistic role (the reserves will interpret "more realistic" as "secondary") to the reserves and to consolidate federal control over the Guard, is substantive reform really possible? If so, how?

James Lacy sums up the problem well:

To be sure, the political obstacles to any serious reexamination of the reserve side of the Total Force are imposing. Located in some five thousand separate communities, reserve units form part of an "intricate and subtle political chain that laces the country, running through village council rooms, county courthouses, and state capitols to Congress and the White House . . . ." Few have been eager to take on that chain in the past, given its recognized clout. Yet, bipartisan reappraisals have been undertaken in other politically thorny thickets, and with occasional success; perhaps a special blue-ribbon commission could at least begin to tackle the job of revising reserve force structure.[2]

The reserves have often charged that the Army plays politics to benefit the active Army at the expense of the reserves. At its 1991 convention, for instance, the Reserve Officers Association asserted that it "is not satisfied that the mobilization of Guard and Reserve units was handled properly and does not trust Defense Department officials to analyze the problems and suggest changes."[3] The former president of the National Guard's Adjutant Generals' Association has stated that the National Guard has "met the enemy, and the enemy is the Department of Defense."[4] But visions of the Department of Defense's top civilian and military leadership sitting in smoke-filled rooms conspiring against the reserves are farfetched.

Indeed, the facts speak otherwise. To say that career Regular Army officers, except for those who have served at the highest levels, are not keen students of the national political scene is somewhat of an understatement. Professional Army officers are weaned on the concept of supporting a decision once it has been made. Active component soldiers are more interested in the nature of the decision than in the politics behind it. The decision was made to support the Total Force Policy and the Army has dutifully done so. The Army, at least as a matter of policy, has bent over backward to support the Total Force Policy and the reserves at the expense of the readiness of the Total Army and at the expense of total honesty.

Active officers are well aware that they are required to support the Total Force Policy. Consequently, they are circumspect in their public comments about the reserves. Their formal evaluations of the reserve components deliberately sugar coat the reserves' shortcomings; the attitude of those who evaluate the reserves is "how can we honestly report our observations without getting anyone in trouble?" An active Army general officer who candidly voiced his opinion concerning the reserves' lack of readiness, Major General Robert E. Wagner, was forced to issue a public apology (and retired from the Army as a major general).

Moreover, accusing the civilian Department of the Army and Department of Defense leadership of supporting the active component at the expense of the reserve components is disingenuous. Civilian political appointees should accede to their posts with no preconceived biases toward the active Army at all (indeed, John O. Marsh, Jr., secretary of the Army throughout the Reagan administration and now chairman of the Reserve Forces Policy Board, is a retired Army National Guard officer). These civilian officials, more than anyone else, are in positions to observe the entire defense establishment and to make policies and recommendations to benefit the military and the national defense as a whole (indeed, that is their mandate). Further, the Department of Defense civilian leadership understands partisan politics well; the current secretary of defense himself is a former congressman.

In sum, when the reserves perceive the active Army and Department of Defense leadership engaging in political maneuvering to snub the reserves, a less biased observer can just as readily perceive honest public officials who, although surely influenced by their experiences, are proposing policies they sincerely believe will benefit the overall national interest. The wisdom of those policies is a separate issue altogether.

This is not to say, however, that the active Army establishment is blameless for the failure to institute meaningful reserve reform. Would-be reserve reformers have consistently failed to grapple with political realities. Upton, for example, heralded military efficiency over civilian control, but predictably failed to convince civilians of the wisdom of his ideas. Palmer could not convince Congress of the merit of universal military training; the idea simply was politically unpopular. McNamara made a frontal charge on the reserves without first determining the strength of their political defenses.

Moreover, the Army historically has sent the reserves the subliminal message that the active Army comes first, even since the advent of the Total Force Policy. Although there may not be a conscious, systemic effort to snub the reserves, in practice the reserves and the active Army are not equal, unintended though this policy may be. The active Army establishment may be sincere in its support of the Total Force Policy, but the essence of the reserves' criticism, despite its vociferousness and exaggeration, may contain a scintilla of truth. After all, it took Congress to ensure that the Army devoted appropriate funds to the reserves in accordance with the Total Force Policy, and active component assignment and promotion policies discourage Regular Army soldiers from serving in reserve component support assignments.

The fact remains, though, that politics are a big, if not the biggest, factor in the reserve equation. When Congress and the reserves play politics, readiness becomes a secondary consideration. As the active Army's strength spirals downward toward the five-hundred-thousand-soldier level, the nation cannot afford to subordinate the readiness of the Army to any other interests. Moreover, just as inefficiency hurts readiness, by definition inefficiency costs money. As the Army shrinks, we must squeeze every ounce of readiness possible out of each defense dollar.

Ironically, implementing such reform implicates politics, one of the obstacles to readiness that reform must eliminate. Obviously, reform of the reserve components on the scale proposed here will require major congressional action. Reformers of the reserve system must meet the National Guard and Reserve lobbies on their own turf. Playing politics is essential to improving the overall readiness of the Army, especially because the reserves, whose parochial interests do not necessarily include the readiness of the Total Army as the top priority, will haul out their political artillery to fight any such reform.

Politics, however, does not necessarily imply deception. The active Army must convince the reserves that its motives are honest and that it will consider the reserves' concerns and involve the reserves in the reform process. As a first step toward implementing a new reserve system, therefore, the active Army must make the present one work better, encouraging active-reserve integration by assigning top quality active officers and noncommissioned officers to reserve support positions and rewarding those soldiers by promoting them.

But honesty cuts both ways. The active Army must discard its rhetorical obsequiousness toward the reserves and adopt a more realistic, yet politically astute, public viewpoint concerning the potential ineffectiveness of the current reserve system in the post-Cold War world. Then the Army must enlist the aid of savvy politicians to convince Congress (and the reserves) at least to listen to its point of view; a principal reason McNamara failed in his bid to reform the reserves twenty-five years ago was his political naïveté. A broad base of support will be required to counter the predictable charges that the active Army is sacrificing the reserves to save its own skin.

Of course, political change cannot occur overnight, especially when many policy makers are not convinced that change is necessary or even that problems exist. Nonetheless, improving the readiness of the reserves is—or should be—an overwhelming national priority, both for Congress and the executive branch. But getting Congress to act in the overall national interest to the detriment of the local concerns of individual congressmen is never easy; after all, congressmen are elected (and reelected) based on local concerns.

Nonetheless, Congress has managed to avert pork-barrel politics in solving other issues of national concern. One example is the formation of an independent bipartisan commission to recommend military base closings. When the commission made its recommendations regarding which bases to close, congressmen cried, "Not in my district!" Congress nevertheless collectively saw the need for change from a national perspective and acted to implement that change as fairly as possible.

Once Congress is convinced of the need for reform—or even that reform should be studied—it should appoint an independent commission to study the entire reserve system. The commission should be given the charter to examine the system from the *national* perspective, without ignoring local concerns, but refusing to let those local concerns cloud the broader issues.

Any plan for reform must not overlook the fact that the reserves are a force of citizen-soldiers. Certainly, valid domestic concerns must be factored into any equation involving the reserves. For example, eliminating the weekend drill system and nationalizing the manpower pool will affect local communities—such a plan may cost jobs and otherwise have an impact on local economies.[5] However, the primary function of military bases is not to provide economic support to the civilian community, and the same is true of the reserve system.

The Gray Board was somewhat analogous to the base closing commission. Why would a commission similar to the base closing commission succeed where the Gray Board failed? The Gray Board was formed in 1948 to study the nation's reserve system and to make recommendations concerning its future, and indeed it made some controversial recommendations, many of which have been reasserted here. But the Gray Board's recommendations died a political death. The critical difference between the Gray Board and the proposed independent commission is that the Gray Board was commissioned by the *executive* branch; it consisted solely of Defense Department officials. Congress had no prior input into the charter of the Gray Board, and, given the outcome and the political climate of the day, probably saw no need for significant reserve reform.

We elect our legislators to make hard decisions. They made those decisions with regard to closing military bases (or, more aptly, they appointed a commission to make those decisions for them so that they would not have to vote on the matter). They can make similar decisions with regard to the reserves. In the end, the benefits of reserve reform to the nation as a whole will far outweigh the costs to local interests and to the special interests of the Guard and Reserve.

Concededly, many of the reforms of the magnitude proposed here may turn out to be simply infeasible politically. But political infeasibility does not lessen the dire need to air the issues. If politics stand in the way of the enactment of such reforms, the least policy makers can do is discuss the issues openly, so that those who in the end will make decisions, or decline to make decisions, as the case may be, will be fully informed as to the consequences of their actions. A full and open discussion of the shortcomings of the current systems, rather than pretending they do not exist, would be an improvement over the status quo. And such discussion might lay the groundwork for changing the political landscape so that meaningful reform can someday be implemented.

There is, however, a glimmer of hope. Three congressmen, dissenting from the House Armed Services Committee's report on its version of the 1993 Defense Authorization Bill, which declined to reduce Army National Guard strength to the level requested by the president, wrote: "Parochial politics—outside the realm of Republican-Democrat squabbles—have again frustrated the ability of the committee to act in the best interests of the nation. . . . When it came time for the committee to make its choices, politics counted and sound national defense policy did not. . . . Military necessity, not political expediency, must shape the Total Force. Perhaps next year the committee will find the courage to act in the interest of military necessity."[6]

Why three armies? Because of tradition and politics. But are those good enough reasons to continue on the present course? The answer is obvious: tradition and politics have led to a military force that potentially could not be up to its ultimate task.

The Soviet threat has gone away, and our civilian policy makers have chosen to reduce the size of the Army. The Army, however, cannot ignore other threats: the volatile Middle East, the unstable Balkans, and the unpredictable North Koreans, for starters. The Army's mission, therefore, has actually become tougher: it now must be prepared to fight the Saddam Husseins of the world with fewer resources. Greater efficiency is essential to the increased readiness the Army's tougher mission demands.

Am I just a modern-day Emory Upton, touting military efficiency over democracy? No. Upton railed against our entire military structure. He advocated a professional military accountable only to itself and saw no potential benefit whatsoever in a citizen-soldier force. I see great potential for a trained reserve.

We should, however, get what we pay for, especially in this age of fiscal belt tightening. If the federal government is to arm, train, feed, clothe, and pay the National Guard, not to mention count on it to provide a substantial portion of our Army's combat power in time of war, then the National Guard ought to be fully accountable to the federal government. Like Upton, I see the current state control of the majority of our Army's reserves as inefficient. In Uptonian times, such inefficiency was a military inconvenience, for the United States had no reason to engage in military excursions to such exotic locales as Iraq and Panama. Today such inefficiency potentially could cause great damage to the very institutions soldiers are sworn to defend.

Abstract concepts of readiness and efficiency aside, there is a much more concrete and convincing reason to reform the reserve components: integration works. When the U.S. Army Recruiting Command received the mission of recruiting for the Army Reserve, two thousand Army Reserve AGR recruiters were assigned to Recruiting Command. Almost immediately, the number of soldiers who enlisted directly from civilian life into the Army Reserve (i.e., those without prior active Army experience and therefore the hardest to recruit) doubled.[7]

Another example of the benefits of true integration: At the U.S. Army Training Center at Fort Jackson, South Carolina, an active component provisional task force and a reserve component provisional training battalion were formed to integrate Army Reserve training divisions into the post's training program. Rather than simply working under an active component training battalion for two weeks and doing only what the active battalion decides is "safe" for the Reserve battalion to do, the Reserve battalions rotate through and man the same provisional battalion during their annual training periods. Four different Army Reserve training battalions, each conducting a two-week annual training period, train the new soldiers assigned to that provisional battalion during the eight weeks of basic training. For continuity, one active Army drill sergeant, stationed at Fort Jackson and actually assigned to the active component provisional task force, is attached to each platoon in the provisional Reserve battalion. The active task force soldiers (except for the continuity drill sergeants) serve as the evaluators for the Army Reserve training battalion; they are, in a sense, "professional evaluators." This integrated system has produced Army Reserve training battalions that are significantly more capable of accomplishing their mobilization missions than does the standard, nonintegrated pattern of Army Reserve training division annual training periods.

Yet another example of the effectiveness of integration is the U.S. Army Special Operations Command. The Special Operations Command commands all Army special operations forces—including Army Reserve special operations forces, which are not assigned to USARC. To streamline its command structure, the Special Operations Command created two subordinate commands. One is an active Army command that commands all active and

Army Reserve special forces units. The other, the U.S. Army Civil Affairs and Psychological Operations Command, commands all active and Reserve civil affairs and psychological operations units.

Because most of the Army's civil affairs and psychological operations forces are in the reserve components, the Civil Affairs and Psychological Operations Command is a reserve component command. The bugs have not yet been completely worked out of this structure; for example, the few active component units assigned to the Civil Affairs and Psychological Operations Command are sometimes frustrated with the command's lack of responsiveness because the Civil Affairs and Psychological Operations Command headquarters is, of course, only partially staffed except on drill weekends and during annual training.

Nonetheless, the new structure has greatly streamlined the lines of authority within the Special Operations Command and has made the command a more efficient organization. As General Wayne A. Downing, until recently the commander of the Special Operations Command, explained:

> Bringing all [reserve component] special operations forces under [the Special Operations Command] enhanced the command's ability to better identify and use our reserve components' capabilities. Making [the command] the Army's single administrative and training manager for all [reserve component special operations forces] also provided a means of ensuring uniform training and readiness standards and a means of validating those standards. This translates into being able to mobilize the right reserve assets rapidly because we know what units are available, their status of training and readiness. Because we do not need to coordinate the activation of a reserve unit with other Army commands, it takes less time to prepare a [reserve special operations unit] for deployment.[8]

Perhaps the most ironic element of the entire complex issue is that improving the reserves' readiness will accomplish the same thing that the reserves heretofore have used their political clout to achieve. If, through integration, the reserves become thoroughly reliable, combat-ready components of the Total Army, the active Army will gain greater confidence in and respect for the reserves, and the reserves' stature will grow. When the active Army becomes convinced of the reserves' capability, the Army will entrust the reserves with the missions and the resources they are now forced to seek through the back door. In the end, almost everybody will win—the Total Army, the reserve components, and the American taxpayer.

The history of the citizen-soldier is colorful and cannot be discarded. Moreover, the Total Force Policy is theoretically sound: trained and capable reserve units are the best source of rapidly available manpower in the event of a national emergency. But we must take care not to overload the reserves with missions unsuited to the citizen-soldier, and we must also ensure that the reserves are capable of performing the carefully considered missions they are given. We must use the lessons of history and the theoretical framework of the Total Force Policy to create a twenty-first-century reserve force that is a more effective and reliable part of the U.S. Army.

This can be done only by eliminating as many distinctions as possible between the three separate armies, while remembering that differences between a citizen-soldier force and a professional army will always remain and indeed are desirable. The major distinctions between the active Army and the reserves are also the biggest stumbling blocks in the path of reserve readiness: state control of the National Guard, geography, ineffective use of training time, and politics. We must address these basic underlying causes of the reserve components' inability to realize their full potential as truly an integral part of the nation's defense structure.

Yes, the reserves are inextricably linked to their past, and yes, the past cannot be discarded. But our emphasis must shift to the future. We must carefully and deliberately examine the future of the reserves based on twenty-first-century, not eighteenth-century, national policies and interests. The national defense of the United States—and indeed of the Western world—is too critical to be overwhelmed by the history and politics of our reserves.

This book may be a voice in the wilderness, but if it can stimulate at least a modicum of informed discussion and thoughtful consideration of the issues, it will have accomplished its purpose. Having said that, however, the importance of these issues cannot be overemphasized. Indeed, the future shape of the reserve components is one of the most critical issues now facing defense planners. Reform of the reserves is essential. The reforms suggested here make sense—militarily and fiscally. And although political ambushes lurk down the road, a bipartisan effort can and should make such reforms a reality.

APPENDIX

TO comprehend fully the implications of the Army's current reserve system requires an understanding of the Total Army within the overall national defense structure, as well as an understanding of the organization and functions of the reserve components. This appendix is an attempt to describe in a few pages the overall system within which the reserve components operate, as well as to discuss in greater detail information mentioned only briefly in chapter 1. The information here is basic to understanding the military in general and the reserves in particular, and is intended primarily for readers who lack familiarity with the military or the reserves.

The nation's armed forces comprise the Army, Navy, Marine Corps, Air Force, and Coast Guard (the Coast Guard, although it is an armed service, is subordinate in peacetime to the Department of Transportation, not the Department of Defense). The armed forces jointly are charged with executing American military policy. Constitutionally, the president is the commander in chief of the armed forces. His authority flows by statute to the secretary of defense. The secretary of defense directly controls the nation's warfighting forces, the "combatant commands."[1] Combatant commands are either unified commands—those consisting of multiservice forces (the United States Central Command of Desert Storm fame, for example), or specified commands—those with a multiservice mission but comprising subordinate elements primarily from a single service (the United States Forces Command is an example; in fact, with the deactivation of the Strategic Air Command, Forces Command is the only remaining specified command, and it is soon to lose that status) (see table 5). Each unified command has an Army component headquarters, which, as a rule, commands Army forces assigned or attached to the unified command.

The combatant commands are either warfighting commands organized geographically (i.e., they are responsible for operations in designated portions of the world) or supporting commands (i.e., they support the geographic commands by performing a specialized function).

Table 5. The Combatant Commands and Their Army Components

Combatant command and army component	Type	Headquarters	Commander in chief	Orientation (warfighting commands)
United States European Command	Unified (warfighting)	Stuttgart, Germany	Army general [1]	Europe and North Africa
Army component: Seventh Army		Heidelberg, Germany	General [2]	
United States Pacific Command	Unified (warfighting)	Camp H.M. Smith, Hawaii	Navy admiral	Pacific
Army component: U.S. Army Pacific		Fort Shafter, Hawaii	Lieutenant general	
United States Central Command	Unified (warfighting)	McDill Air Force Base, Fla.	Army or Marine general	Southwest Asia
Army component: Third Army		Fort McPherson, Ga.	Lieutenant general	
United States Southern Command	Unified (warfightling)	Quarry Heights, Panama City, Panama	Army general	Central and South America
Army component: U.S. Army South		Fort Clayton, Panama	Major general	
United States Atlantic Command	Unified (warfighting)	Norfolk, Virginia	Navy admiral	Atlantic and Caribbean
Army component: Forces Command [3]		Fort McPherson, Ga.	General	
United States Space Command	Unified (supporting— aerospace warfare)	Peterson Air Force Base, Col.	Air Force general	
Army component: U.S. Army Space Command		Peterson Air Force Base, Col.	Brigadier general	

Table 5. Continued

Combatant command and army component	Type	Headquarters	Commander in chief	Orientation (warfighting commands)
United States Transportation Command	Unified (supporting—strategic airlift and sealift)	Scott Air Force Base, Ill.	Air Force general	
Army component: Military Traffic Management Command		Falls Church, Va.	Major general	
United States Special Operations Command	Unified (supporting—provision of Special Operations Forces)	McDill Air Force Base, Fla.	Army general	
Army component: U.S. Army Special Operations Command		Fort Bragg, N.C.	Lieutenant general	
United States Forces Command[4]	Specified (warfighting—territorial defense—and supporting—Strategic Reserve)	Fort McPherson, Ga.	Army general	

1. Also serves as supreme allied commander, Europe, or commander in chief of all NATO forces, with headquarters in Mons, Belgium.
2. Also serves as commander in chief, U.S. Army Europe, a major Army command.
3. Also a specified command. Forces Command assigned as the Army component for planning purposes only until activated by the Department of Defense for operations.
4. Also a major command of the Army. Forces Command's joint missions are to be assumed by the United States Atlantic Command, divesting Forces Command of its specified status.

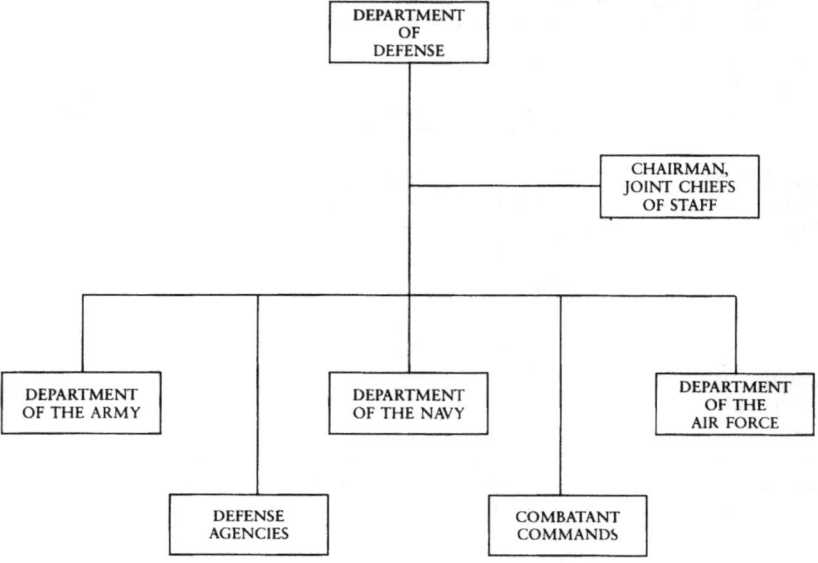

Figure 6. Department of Defense Structure

Commanders in chief of the combatant commands are directly subordinate to the secretary of defense.

Although the chairman of the Joint Chiefs of Staff officially is not part of this operational chain of command, he plays an important role in the national defense establishment. The chairman is the senior U.S. military officer and is the president's principal military adviser. The chairman assists the president and the secretary of defense in providing strategic direction to the armed services and prepares joint operational and logistical plans.[2] Additionally, the chairman oversees the activities of the combatant commands and serves as the conduit between the combatant commanders in chief and the secretary of defense and the president (but again, the chairman of the Joint Chiefs of Staff is not in the chain of command).

In addition to controlling the combatant commands directly, the secretary of defense directly controls the military departments (see fig. 6). The Department of the Army is headed by the secretary of the Army. The top uniformed soldier is the chief of staff of the Army. The Department of the Army's function is to train, administer, equip, and support ground forces assigned to, or assignable to, the combatant commands. As a whole, therefore, the Army is not a warfighting command.

APPENDIX

The Army's tactical units (i.e., those that will be provided to the combatant commands, as opposed to training units, procurement organizations, and garrisons that staff Army installations, stateside maintenance depots, service schools, etc.) are categorized by their function on the battlefield. Combat units are those whose missions involve actually fighting the battle, such as infantry, armor, and artillery units. Combat support units—engineer, military police, and signal units, for example—provide operational assistance to the combat forces on the battlefield. Combat service support units provide maintenance, supply, medical, transportation, and other battlefield services.[3] The various types of tactical units are described in table 6. Additionally, table 7 provides a brief snapshot of the people who lead those units.

In addition to its active component (that is, its active-duty personnel), each of the armed services has one or more reserve components. The nation's reserves, therefore, consist of the Army National Guard, the Army Reserve, the Naval Reserve, the Marine Corps Reserve, the Air National Guard, the Air Force Reserve, and the Coast Guard Reserve.[4] The Army's reserve components are the Army National Guard, a force that has its roots in the colonial militia and is jointly controlled by the Army and the states, and the United States Army Reserve, a fully federal reserve force.

At the Department of Defense level, the assistant secretary of defense (reserve affairs) is the principal assistant to the secretary of defense for reserve matters and is vested with overall responsibility for all of the military services' reserve components. Congress did not create the office of assistant secretary of defense for reserve affairs until 1984. The assistant secretary of defense for reserve affairs is appointed by the president with the advice and consent of the Senate.

The Reserve Forces Policy Board, created by statute, acts through the assistant secretary of defense for reserve affairs as the principal policy adviser to the secretary of defense on reserve matters.[5] The Reserve Forces Policy Board comprises a civilian chairman, the service assistant secretaries for manpower and reserve affairs, and various general and flag officers, both active and reserve, from each service.

At the Department of the Army level, the assistant secretary of the Army (manpower and reserve affairs) oversees the Army's reserve components. The assistant secretary of the Army for manpower and reserve affairs supervises and directs the formulation, execution, and review of Army policies, plans, and programs concerning the reserve components.[6] The deputy assistant secretary of the Army for reserve forces and mobilization is directly subordinate to the assistant secretary of the Army for manpower and reserve affairs and is the highest civilian De-

Table 6. Major Units of the Army Force Structure

Level of organization	Description
Echelons above corps	Elements such as theater armies and major commands that have authority over corps.
Corps	The Army's four active corps command, control, and support several divisions. Each corps has both tactical and logistical responsibilities and provides auxiliary combat arms and services such as artillery, corps support, air defense, aviation, and engineering, intelligence, and military police to its divisions.
Division	The Army's twenty-four divisions serve as its major tactical units. Each generally consists of three brigades, although they can control more, and nine to ten maneuver battalions. A division combines the combat arms and services required for sustained combat. Some of the forces to support and sustain division operations are provided by the corps and echelons above corps. *Heavy divisions* include the armored and mechanized infantry divisions, each with about 17,000 soldiers. *Light infantry divisions* have about 10,000 personnel, possess lighter equipment than the heavy divisions and are designed to deploy rapidly. The 82d Airborne Division and the 101st Airborne Division (Air Assault) are slightly larger light divisions with special air capabilities (parachute for the 82d, helicopter for the 101st).
Brigade	A division generally consists of three brigades, each consisting of two to five combat battalions. A brigade comprises 3,000-4,500 soldiers.
Separate brigade	Separate brigades are independent units that possess the support forces necessary to sustain their operations. The corps commander can use them either in the corps battle or in independent operations. An armored cavalry regiment is a specialized brigade-sized independent unit that is organized to conduct corps reconnaissance and security operations.

Table 6. Continued

Level of organization	Description
Battalion	Battalions normally consist of five companies and are tactically and administratively self-sufficient. They vary in size, consisting of between 550 and 825 soldiers, depending on their type, and are capable of independent operations of limited duration and scope. An armored or air cavalry battalion-sized unit is called a squadron.
Company	A company consists of three to four platoons and contains from 100 to 180 soldiers. Companies generally do not operate independently. An artillery company-sized unit is called a battery; a cavalry company-sized unit is called a troop.
Platoon	A platoon consists of three to four squads or four to five armored vehicles. There are from twenty to forty-five soldiers in a platoon.
Squad	There are from eight to twelve personnel in a squad. The basis or organization for an armored unit is the tank crew rather than the squad. Other units, such as artillery, that are organized around their weapons systems rather than around their personnel also have crews rather than squads as their lowest level units.

Source: GAO, *Army Force Structure: Lessons to Apply in Structuring Tomorrow's Army* (Washington, D.C., 1990).

partment of the Army official who devotes his time solely to reserve matters.

The Army's analogue to the Department of Defense's Reserve Forces Policy Board is the Reserve Forces Policy Committee, organized in the office of the secretary of the Army. The committee, consisting of five senior officers from each of the Army's three components, reviews and comments on major policy matters affecting both the reserves and the mobilization preparedness of the Army.[7]

As explained in chapter 1, the Army, under the Total Force Policy, has made deliberate decisions to place much of its capability in the reserve components. To increase its deterrence capability, for example,

Table 7. Selected Information Concerning Army Leaders and Commanders

Unit	Number of soldiers	Rank of commander or leader	Age of commander or leader	Years of service of commander or leader
Squad or vehicle crew	8-12	Sergeant or staff sergeant	22-28	4-8
Platoon	20-45	Second or first lieutenant	22-26	0-3
Company/ battery/ troop	100-180	Captain[1]	26-30	4-8
Battalion/ squadron	550-825	Lieutenant colonel	36-46 (average 40)	15-21 (average 17)
Brigade	3,000-4,500	Colonel[2]	40-50 (average 44)	18-29 (average 21)
Division	10,000-17,000	Major general	48-54 (average 50)	27-32 (average 29)

1. Aviation companies and air cavalry troops are normally commanded by majors who are commensurately older and have more years of service.
2. Separate brigades (those not part of a division), especially those in the reserve components, are often commanded by brigadier generals.

Source: Robert L. Goldich, *The Army's Roundout Concept after the Persian Gulf War.* (Washington, D.C.: Congressional Research Service, 1991).

the Army has sacrificed significant support capability in the active component to maintain active combat forces; currently, more than two-thirds of the Army's tactical support capability is in the reserve components.[8] The reserves contain all of the Army's chemical brigades, all of its water supply battalions, all of its separate judge advocate general units, 97 percent of its civil affairs units, 87 percent of its public affairs units, and 86 percent of its medical brigades.[9] The implications of these statistics are both obvious and significant (see table 8).

The reserve components are divided into the Ready Reserve, the Standby Reserve, and the Retired Reserve. The Ready Reserve is further divided into the Selected Reserve, the Individual Ready Reserve, and the Inactive National Guard. The Selected Reserve is the most readily mobilized and consists of reserve units, both Army National Guard and Army Reserve; individual mobilization augmentees; and full-time support personnel. As of December 31, 1991, the Selected Reserve consisted of 443,253 soldiers in Army National Guard units, 305,572 soldiers in Army Reserve units, and 13,985 individual mobilization augmentees.[10] As of September 30, 1992, there were 25,024 AGR personnel in the Army National Guard and another 13,146 in the Army Reserve, and 28,138 military technicians assigned to the Army National Guard and 8,112 to the Army Reserve.[11] At the end of fiscal year 1992, there were 974 active Army full-time support soldiers, 109 supporting the Army Guard and 865 supporting the Army Reserve.[12]

The Army National Guard, including the Inactive National Guard, consisted of 451,584 soldiers as of December 1991.[13] The Army National Guard provides 44 percent of the Total Army's combat units, including 42 percent of its combat divisions and 31 percent of its tactical support units.[14] Some of the Army's capabilities reside only in the Army National Guard: the Guard provides all of the Army's heavy helicopter units, infantry scout groups, and TOW (tube launched, optically tracked, wire command link anti-tank guided missile) light anti-tank battalions.[15]

The Army Reserve also contributes heavily to the Total Army's capability, providing 17 percent of its nondivisional combat units, 42 percent of its tactical support units, and 27 percent of its general support units.[16] As of September 30, 1991, the Reserve, including the Individual Ready Reserve and the Standy Reserve, comprised 677,765 soldiers.[17]

The Individual Ready Reserve and the Inactive National Guard cannot be mobilized absent, at a minimum, a presidential declaration of emergency. The Inactive National Guard consists of Guard soldiers who are temporarily unable to train with their Guard units. Inactive Guard soldiers are transferred to the Individual Ready Reserve if they do not resume their active Guard status within a year after transfer to the Inactive National Guard. Individual Ready Reserve and Inactive National Guard soldiers are not assigned to units and are required to muster once annually. There were a total of 371,040 Individual Ready Reserve and 8,331 Inactive National Guard soldiers at the end of fiscal year 1991.[18]

The Standby Reserve is composed of Reservists in an inactive sta-

Table 8. The Reserve Components' Contribution to the Total Army

Unit Types	Number Units Army National Guard	Number Units Army Reserve	Combined Percent of Total Army
Infantry Divisions	5	0	100
Heavy Helicopter Units	3	0	100
Chemical Brigades	1	3	100
Water Supply Battalions	2	3	100
Rail Battalions	0	2	100
Training Divisions	0	12	100
Training Brigades	0	3	100
Judge Advocate General Units	4	137	100
Enemy Prisoner of War Brigades	0	1	100
Theater Area Support Groups	0	1	100
Civil Affairs Units	0	36	97
Petroleum Support Battalions	7	6	93
Public Affairs Units	32	26	87
Medical Brigades	3	9	86
Chemical Battalions	2	9	85
Separate Brigades	10	1	79
Medical Groups	3	7	77
Maintenance Battalions	21	17	73
Corps Support Groups	4	15	73
Psychological Operations Units	0	36	72
Motor Battalions	10	14	71
Engineer Battalions (Combat Heavy)	14	17	70
Ordnance Battalions	1	5	67
Petroleum Groups	0	2	67
Theater Defense Brigades	3	1	67
Hospitals	21	50	66
Engineer Battalions (Combat)	42	16	64
Field Artillery Battalions	95	11	63
Aviation Assault Battalions	12	5	61
Terminal Battalions	0	4	57
Military Police Battalions	9	7	55
Military Police Brigades	4	2	55
Medium Helicopter Battalions	4	2	55
Corps Support Commands	2	2	50

Table 8. Continued

Unit Types	Number Units Army National Guard	Number Units Army Reserve	Combined Percent of Total Army
Armor Divisions	2	0	50
Area Support Groups	9	8	49
Attack Helicopter Battalions	21	3	45
Special Forces Groups	2	2	44
Aviation Brigades	15	0	41
Theater Army Area Commands	0	2	40
Armored Cavalry Regiments	2	0	40
Signal Battalions	31	4	40
Military Intelligence Battalions	7	15	39
Air Defense Brigades	3	0	33
Air Defense Battalions	17	0	33
Mechanized Infantry Divisions	2	0	29
Engineer Battalions (Topographical)	1	0	25
Light Infantry Divisions	1	0	20

Source: Office of the Secretary of Defense, *Reserve component Programs—FY 1992: Report of the Reserve Forces Policy Board* (Washington, D.C., 1993).

Note: Data as of September 30, 1992.

tus, such as key federal employees or individuals temporarily precluded from serving in the Ready Reserve because of disability or hardship. The Standby Reserve can be mobilized only in time of war or national emergency and then only when there are insufficient Ready Reserves. With the creation of the Selected Reserve, the Individual Ready Reserve has largely assumed the role the Standby Reserve was intended to play. The Standby Reserve, as of December 31, 1991, consisted of only 1,153 personnel.[19]

The Retired Reserve consists of all reservists receiving retired pay (retired reservists become eligible to receive retired pay at age sixty); reservists who, although otherwise eligible for retired pay, have not reached age sixty, have not elected discharge from the Army, and are not voluntary members of the Ready or Standby Reserve; and enlisted soldiers who retired from active service with either the active component or the reserve components with at least twenty but less than thirty years of service. The latter category of retiree may be transferred to the appropriate retired list after completion of thirty total years of service (i.e., active service plus service in the Retired Reserve). Twenty-year enlisted members of the Retired Reserve may be ordered to active duty at any time.[20]

Finally, it is interesting to note that the militia still exists and that it comprises more than just the National Guard. Today the militia, by law, consists of both the organized militia, primarily the National Guard; and the unorganized militia, which consists of all able-bodied male citizens at least seventeen years of age, with certain exceptions (examples of the exceptions are the vice-president, federal and state officials, and members of the merchant marine.)[21]

GLOSSARY

Active Guard/Reserve: National Guard or Reserve soldier on active duty for more than 180 days.
Active component: The Regular Army.
AGR: Active Guard/Reserve.
Air cavalry: Helicopter units specially organized and trained for reconnaissance and security operations, normally found in light forces.
AirLand Battle: The U.S. Army's warfighting doctrine.
ARCOM: Army Reserve command; two-star headquarters that commands Army Reserve units in a given geographical area.
Armored cavalry: Armored units specially organized and trained for reconnaissance and security operations.
Armory: The building that houses an Army National Guard unit.
Army component headquarters: The senior Army headquarters that generally commands all Army forces assigned or attached to a command composed of more than one service.
Army Training and Evaluation Program: The Army program that prescribes the missions a unit must be capable of performing and the standards for performing those missions.
Civil affairs: Activities that embrace the relationship between military forces and the civil authorities and population in a country or area, including performance by the military forces of certain functions or the exercise of certain authority normally the responsibility of the local government.
Combat service support units: Units whose primary missions are to perform the essential logistic functions, activities, and tasks necessary to sustain all elements of an operating force.
Combat ready: Capable of performing the missions or functions for which a unit is organized or designed.
Combat support units: Units whose primary missions are to provide operational assistance for combat units.
Combat units: Units whose primary missions are to engage the enemy by means of fire and/or maneuver.

Combatant command: A unified or specified command.

Communications zone: The rear area of the theater of operations, behind the corps, where theater army support forces are stationed.

Contingency: An emergency involving military forces caused by natural disasters, terrorists, subversives, or required military operations. Contingencies require rapid response.

CONUSA: Continental United States Army, an active component organization organized to supervise the reserve components in a designated geographic area.

Deployment: Movement of forces from peacetime positions to a theater of operations.

Employment: The use of military forces on the battlefield.

Federal recognition: The process by which a National Guard officer is commissioned or promoted as an officer in the reserve components of the U.S. Army.

Federalization: The ordering of National Guard units into federal service.

Force mix: The proportion of active forces in relation to reserve forces in the Army force structure.

Force structure: Numbers, size, and composition of units that make up the Army.

Forward deployed: Stationed outside the continental United States.

G-3: The general staff officer with primary responsibility for operations, plans, and training.

Interoperability: The ability of systems, units, or forces to provide services to and accept services from other systems, units, or forces and to use the services so exchanged to enable them to operate effectively together; for example, the ability of one item of communications equipment to "talk" to another item of communications equipment.

Judge advocate general officer: A military lawyer.

Light forces: Units that are not mechanized, are organized with a minimum of organic transportation, and rely principally on the individual soldier to provide the bulk of their combat power.

Mechanized infantry: Infantry forces equipped with armored personnel carriers or infantry fighting vehicles.

Military occupational specialty: The individual function a soldier is trained to perform; a soldier's "job."

Mission essential task list: A list of tasks, derived from the Army Training and Evaluation Program, that a unit commander deems most important to the accomplishment of the unit's wartime mission and that require the greatest training emphasis.

GLOSSARY

Noncommissioned officer: An enlisted soldier in the grade of corporal or above.

Officer efficiency report: The document by which an officer's performance and potential are evaluated by the officer's superiors, prepared at least annually.

Operation plan: A plan for a single or series of connected military operations to be carried out simultaneously or in succession; in short, a war plan.

Psychological operations: Planned psychological activities directed to enemy, friendly, and neutral audiences in order to influence attitudes and behavior affecting the achievement of political and military objectives.

Readiness: See *combat ready.*

Reserve center: The building that houses an Army Reserve unit.

Reserve components: The Army National Guard and the U.S. Army Reserve.

Secure communications: Encrypted radio or other telecommunication signals that require special equipment to decrypt.

Separate brigade: A brigade that is not part of a division.

Special operations: Operations conducted by specially trained, equipped, and organized forces in pursuit of national military, political, economic, or psychological objectives. These operations may support conventional operations or may be prosecuted independently.

Specified command: A command that has a broad continuing mission and that is established and so designated by the president through the secretary of defense. It normally is composed of forces from but one service.

State area command: A National Guard headquarters, commanded by the state adjutant general, formed in each state to control mobilization of the state's National Guard forces.

Tactical: Organized for, or pertaining to, a combat mission.

Theater army: A unit, larger than a corps, responsible for the administrative control of all Army forces in a theater of operations.

Theater of operations: That portion of the world within which military operations are conducted.

Time in grade: The length of time a soldier has held his or her current rank.

Time in service: The length of time a soldier has been in the Army.

Unified command: A command with a broad continuing mission under a single commander and composed of significant assigned components of two or more services, and which is established and so designated by the president through the secretary of defense with the advice and assistance of the joint chiefs of staff, or, when so authorized by the joint chiefs of staff, by a commander of an existing unified command established by the president.

United States property and fiscal officer: An Army or Air National Guard officer on extended active duty who is responsible for a state National Guard's expenditure of federal funds.

USARC: U.S. Army Reserve Command; a major subordinate command of Forces Command that commands all Army Reserve units, less special operations units, in the continental United States.

Wartime gaining command: The unit or headquarters to which, according to operation plans, a reserve unit will be assigned or attached upon mobilization.

NOTES

1. The Three-Army System

1. James L. Lacy, "Whither the All-Volunteer Force?" *Yale Law & Policy Review* 5 (Fall/Winter 1986): 57.
2. General Colin L. Powell, *National Military Strategy of the United States* (Washington, D.C., 1992), 4.
3. The 2d Armored Division, 3d Armored Division, 8th Infantry Division (Mechanized), and 9th Infantry Division (Motorized) have recently been inactivated.
4. Actually, the commander in chief of Forces Command wears three hats, also serving, for planning purposes, as the Army omponent commander for U.S. Atlantic Command (see the Appendix). Forces Command is to cede its territorial defense responsibility to the U.S. Atlantic Command and will then lose its status as a specified command.
5. Because the states also control the Air National Guard, some adjutants general are Air National Guard officers.
6. Peter A. Fish, "The Constitution and the Training of National Guardsmen: Can State Governors Prevent Uncle Sam from Sending the Guard to Central America?" *Journal of Law and Politics* 4 (Winter 1988): 625, citing the congressional testimony of Assistant Secretary of Defense (Reserve Affairs) James H. Webb, Jr.
7. U.S. Army Training Board, *Training and Organization of the Reserve Components: A Reference Text for Total Force Trainers, 1987-88* (Ft. Monroe, Va., 1988), 17 (hereafter cited as *Training and Organization of the Reserve Components*).
8. Lt. Col. Sol Gordon, ed., *1986 National Guard Almanac* (Washington, D.C.: Uniformed Services Almanac, 1986), 76.
9. 10 U.S.C. § 3040 (1988).
10. Colonel David S. Ritterpusch, *The Army Reserve and U.S. National Defense Strategy* (U.S. Army War College Military Studies Program paper, 1988), 37-38.
11. *Defense Authorization Act for Fiscal Year 1991*, Pub. L. 101-510, sec. 903, 104 Stat. 1620 (1990).
12. Office of the Secretary of Defense, *Reserve Component Programs, Fiscal Year 1990: Report of the Reserve Forces Policy Board* (Washington,

D.C., 1991), 80 (hereafter cited as *1990 Reserve Forces Policy Board Report*).

13. Association of the United States Army, *The Active and Reserve Components, Partners in the Total Army* (Arlington, Va., 1989), 14 (hereafter cited as *Partners in the Total Army*).

14. Department of the Army, CAPSTONE *Program*, Army Regulation 11-30 (Washington, D.C., 1985), 4.

15. Ibid., 5.

16. *Training and Organization of the Reserve Components*, 41-42.

17. Army Regulation 11-30, 4.

18. 10 U.S.C. § 331, 332, 3500 (1988).

19. 10 U.S.C. § 673b (1988).

20. 10 U.S.C. § 673 (1988).

21. 10 U.S.C. § 672 (1988).

22. Martin Binkin and William W. Kaufmann, *U.S. Army Guard and Reserve: Rhetoric, Realities, Risks* (Washington, D.C.: Brookings Institution, 1989), 62.

23. Major General Roger W. Sandler, "When the Call Came Reserve Was Ready," *Army*, Oct. 1991, 103.

24. Binkin and Kaufmann, *U.S. Army Guard and Reserve*, 29.

25. Ibid.

26. Ibid.

27. Ibid., 30-31.

2. The Evolution of Three Armies

1. John K. Mahon, *History of the Militia and the National Guard* (New York: Macmillan, 1983), 15. I have drawn extensively upon Professor Mahon's book throughout this chapter. Rather than citing each individual historical proposition I have borrowed from Mahon, I have ascribed to him only his major ideas and direct quotes.

2. Ibid., 19.

3. Alan Hirsch, "The Militia Clauses of the Constitution and the National Guard," *University of Cincinnati Law Review* 56, no. 3 (1988): 923.

4. Mahon, *History of the Militia and the National Guard*, 66.

5. Ibid., 68.

6. Frederick B. Wiener, "The Militia Clauses of the Constitution," *Harvard Law Review* 54 (Dec. 1940): 189.

7. Russell F. Weigley, *History of the United States Army* (New York: Macmillan, 1967), 120.

8. Mahon, *History of the Militia and the National Guard*, 69.

9. Wiener, "The Militia Clauses," 190.

10. Weigley, *History of the United States Army*, 183.

11. Wiener, "The Militia Clauses," 190.

12. Richard B. Crossland and James T. Currie, *Twice the Citizen: A History of the Army Reserve, 1908-1983* (Washington, D.C.: Office of the Chief, Army Reserve, 1984), 9.

13. Mahon, *History of the Militia and the National Guard*, 104.
14. Wiener, "The Militia Clauses," 190.
15. Frederick Porter Todd, quoted in Mahon, *History of the Militia and the National Guard*, 106.
16. Mahon, *History of the Militia and the National Guard*, 110.
17. Ibid.
18. Ibid., 119.
19. Weigley, *History of the United States Army*, 278.
20. Emory Upton, *The Military Policy of the United States* (Washington, D.C., 1907), xiii.
21. Weigley, *History of the United States Army*, 278.
22. Upton, *The Military Policy of the United States*, xiv.
23. Weigley, *History of the United States Army*, 281.
24. Mahon, *History of the Militia and the National Guard*, 130.
25. Weigley, *History of the United States Army*, 306.
26. Ibid., 307.
27. Wiener, "The Militia Clauses," 194-95.
28. Ibid., 195.
29. U.S. House of Representatives, 57th Cong., 1st sess., 1902, H. Rept. 1094, 22-23, quoted in Wiener, "The Militia Clauses," 195.
30. Mahon, *History of the Militia and the National Guard*, 142.
31. Weigley, *History of the United States Army*, 344-45.
32. Ibid., 345.
33. Mahon, *History of the Militia and the National Guard*, 153.
34. Ibid.
35. Crossland and Currie, *Twice the Citizen*, 31.
36. Mahon, *History of the Militia and the National Guard*, 158.
37. Ibid., 163.
38. Ibid., 170.
39. Crossland and Currie, *Twice the Citizen*, 36-37.
40. Wiener, "The Militia Clauses," 208.
41. Ibid., 209.
42. Ibid., 208.
43. Crossland and Currie, *Twice the Citizen*, 37, 45-59, 62.
44. Mahon, *History of the Militia and the National Guard*, 178-79.
45. Crossland and Currie, *Twice the Citizen*, 65.
46. Mahon, *History of the Militia and the National Guard*, 181.
47. Jim Dan Hill, *The Minuteman in Peace and War* (Harrisburg, Pa.: Stackpole, 1964), 391.
48. Ibid.
49. Mahon, *History of the Militia and the National Guard*, 186; Crossland and Currie, *Twice the Citizen*, 78.
50. Crossland and Currie, *Twice the Citizen*, 66-67.
51. Ibid., 85.
52. Committee on Civilian Components, *Reserve Forces for National Security: Report to the Secretary of Defense* (Washington, D.C., 1948), 10 (hereafter cited as *Gray Board Report*).

53. Mahon, *History of the Militia and the National Guard*, 201, quoting the *National Guardsman* (Sept. and Dec. 1948).
54. Ibid., 201-02.
55. Lacy, "Whither the All-Volunteer Force?" 61.
56. Crossland and Currie, *Twice the Citizen*, 99.
57. *Partners in the Total Army*, 9.
58. Crossland and Currie, *Twice the Citizen*, 86-87.
59. Mahon, *History of the Militia and the National Guard*, 210-11.
60. Ibid., 215.
61. Ibid., 218-19.
62. Crossland and Currie, *Twice the Citizen*, 132.
63. Ibid., 145-46.
64. *Partners in the Total Army*, 9.
65. Crossland and Currie, *Twice the Citizen*, 207.
66. General Bruce Palmer, Jr., *The 25-Year War: America's Military Role in Vietnam* (Lexington: Univ. Press of Kentucky, 1984), 175.
67. Binkin and Kaufmann, *U.S. Army Guard and Reserve*, 62.
68. Ibid., 41, citing Gus C. Lee and Geoffrey Y. Parker, *Ending the Draft: The Story of the All Volunteer Force* (Alexandria, Va.: Human Resources Research Organization, 1977).
69. President George Bush, quoted in Office of the Secretary of Defense, *Reserve Component Programs, Fiscal Year 1988: Report of the Reserve Forces Policy Board* (Washington, D.C., 1989), i (hereafter cited as *1988 Reserve Forces Policy Board Report*).
70. *Partners in the Total Army*, 11.
71. Binkin and Kaufmann, *U.S. Army Guard and Reserve*, 1, 13. Information concerning these European war plans were obtained from an unclassified source, so obviously they are no longer current. They nonetheless illustrate the point.
72. *Partners in the Total Army*, 10.
73. 10 U.S.C. § 672(f) (1988).

3. Systemic Disconnects in the Total Army Circuit

1. Department of the Army, *Operations*, Field Manual 100-5 (Washington, D.C., 1986), 1.
2. Liz Galtney, "The Sad State of Weekend Warriors," *U.S. News & World Report*, Sept. 25, 1989, 28.
3. Martin Binkin and William W. Kaufmann, *U.S. Army Guard and Reserve*, 94.
4. Caleb Baker, "Reserve Forces Examine Larger Role in U.S. Military," *Defense News*, Aug. 21, 1989, 16.
5. Office of the Secretary of Defense, *Reserve Component Programs—FY 1992: Report of the Reserve Forces Policy Board* (Washington, D.C.,

NOTES TO PAGES 51-65

1993), 101 (hereafter cited as *1992 Reserve Forces Policy Board Report*).

6. General Accounting Office, *Army Reserve Components: Minimum Essential Equipment for Training Has Not Been Effectively Managed* (Washington, D.C., 1990), 6.

7. General Accounting Office, *National Guard: Peacetime Training Did Not Adequately Prepare Combat Brigades for Gulf War* (Washington, D.C., 1991), 19-20 (hereafter cited as GAO *Readiness Report*). The Reserve Component Automation System, currently under development, is designed to bridge the gap between the components' different systems. Nonetheless, the Army National Guard will be required to use the active Army systems upon mobilization.

8. Ritterpusch, *The Army Reserve*, 13.

9. Office of the Chief, Army Reserve, *The Chief, Army Reserve's Long-Range Plan, 1990-2020* (Washington, D.C., 1990), 3-4.

10. Lacy, "Whither the All-Volunteer Force?" 59.

11. General Accounting Office, *Army Training: Management Initiatives Needed to Enhance Reservists' Training* (Washington, D.C., 1989), 20 (hereafter cited as GAO *Training Report*).

12. Ibid., 36.

13. Office of the Secretary of Defense, Reserve Component Programs: Report of the 1989 Reserve Forces Policy Board, Fiscal Year 1989 (Washington, D.C., 1990), 80 (hereafter cited as *1989 Reserve Forces Policy Board Report*).

14. The governors eventually lost their challenge in the United States Supreme Court. See *Perpich v. United States Department of Defense*, 110 S. Ct. 2418 (1990).

15. "Force Structure: Active Army," *Army*, April 1992, 47.

16. General John A. Wickham, Jr., *Chief of Staff's White Paper on the Light Infantry Division, Army of Excellence* (Washington, D.C., 1984), 1.

17. *Gray Board Report*, 12.

18. Ibid.

19. Department of the Army, *Training the Force*, Field Manual 25-100 (Washington, D.C., 1988).

20. Lacy, "Whither the All-Volunteer Force?" 57.

21. General Accounting Office, *Army Training: Evaluations of Units' Proficiency Are Not Always Reliable* (Washington, D.C., 1991), 38-39.

22. Ibid., 37.

23. Ibid., 39.

24. William V. Kennedy, "The Reserve Dilemma," *Washington Post*, Dec. 26, 1990, p. A24.

25. Lacy, "Whither the All-Volunteer Force?" 60.

26. General Accounting Office, *Reserve Components: Opportunities to Improve National Guard and Reserve Policies and Programs* (Washington, D.C., 1988), 29-30.

27. *Training and Organization of the Reserve Components*, 10.

28. *1990 Reserve Forces Policy Board Report*, 66.

29. *Training and Organization of the Reserve Components*, 10.
30. General Accounting Office, *Individual Ready Reserve: Army Needs to Make More Effective Use of Limited Training Funds* (Washington, D.C., 1990), 13.
31. Ibid., 14.
32. Ibid., 10.
33. Binkin and Kaufmann, *U.S. Army Guard and Reserve*, 76.
34. Office of the Secretary of Defense, *Reserve Component Programs, Fiscal Year 1991: Report of the Reserve Forces Policy Board* (Washington, D.C., 1992), 65 (hereafter cited as *1991 Reserve Forces Policy Board Report*).
35. Robert L. Goldich, *Persian Gulf War: U.S. Reserve Callup and Reliance on the Reserves* (Washington, D.C.: Congressional Research Service, 1991), 13. Goldich concludes that the success of the Desert Storm Individual Ready Reserve mobilization is perhaps more attributable to the quality of soldiers the active Army has been able to recruit over the past few years.
36. Bennie J. Wilson III and James R. Engelage, "Pretrained Individual Manpower: Albatross or Phoenix?" in *The Guard and Reserve in the Total Force*, ed. Bennie J. Wilson III (Washington, D.C.: National Defense Univ. Press, 1985), 128.
37. Arthur L. Moxon, "U.S. Reserve Forces: The Achilles' Heel of the All-Volunteer Force?" in Wilson, ed., *The Guard and Reserve in the Total Force*, 102.

4. Geography, Time, and Other Readiness Detractors

1. *Training and Organization of the Reserve Components*, 4.
2. Ibid.
3. *Partners in the Total Army*, 19.
4. *1988 Reserve Forces Policy Board Report*, 84.
5. Captain German J. Velez, "48 Hours: Fighting the Reserve Component Battle," *Infantry*, Nov.-Dec. 1989, 17.
6. WESTAT, Inc., *Survey of United States Army Reserve (USAR) Troop Program Units (TPU) Soldiers*, Technical Report, Tabulation of Questionnaire Responses, Junior Enlisted Personnel, 447; see also Lieutenant Colonel Terry White and Major Jim Simpson, "Why Soldiers 'Really' Leave the Army Reserve: The WESTAT Survey," *Military Review*, Feb. 1990, 77.
7. Binkin and Kaufmann, *U.S. Army Guard and Reserve*, 99, citing a 1987 U.S. Army Training Board study.
8. A new Army program dubbed "Bold Shift" has sought to refocus reserve component combat unit training at a lower level. At least according to one opinion, however, Bold Shift has not necessarily been an unqualified success. See "Reserve Training: Bold Shift Just B.S.," *Army Times*, Oct. 12, 1992, 36.

9. Bernard E. Trainor, "Pentagon Worried by Use of Reserves," *New York Times*, July 16, 1989, p. 1.
10. Galtney, "The Sad State of Weekend Warriors," 28.
11. *1990 Reserve Forces Policy Board Report*, 68.
12. Soldiers who fill slots for which they are not qualified take the skill qualification test in their own military occupational specialty. Thus, an infantryman assigned as a supply sergeant takes the infantry skill qualification test. Admittedly, the calculation is overly simplistic because the it assumes that all of the unqualified 21 percent is included in the 65 percent that passed the skill qualification test.
13. Ritterpusch, *The Army Reserve*, 12.
14. GAO *Training Report*, 40.
15. *1989 Reserve Forces Policy Board Report*, 80.
16. GAO *Training Report*, 24.
17. Ibid., 33.
18. *Training and Organization of the Reserve Components*, 5.
19. *Partners in the Total Army*, 13, citing the Army's Reserve Component Training Development Action Plan promulgated in 1989.
20. Galtney, "The Sad State of Weekend Warriors," 29.
21. L. James Binder, "Vital to Army that Reductions be Managed," *Army*, Feb. 1990, 17.
22. John O. Marsh, Jr., "The Army in the Total Force Structure: Readiness, Reserves and Demographics," speech delivered March 28, 1990, summarized in Association of the United States Army, Background Briefs, *The Future U.S. Army—Outlook for the 1990s* (Arlington, Va., 1990), No. 12. Marsh is the chairman of the Reserve Forces Policy Board and a former secretary of the Army.
23. *Training and Organization of the Reserve Components*, 5.
24. GAO *Readiness Report*, 13.
25. *1990 Reserve Forces Policy Board Report*, 68. *Qualified* means that a soldier has received the training necessary to be awarded a particular military occupational specialty. The term should not be confused with proficiency in that military occupational specialty as measured by the skill qualification test.
26. Ibid., 173.
27. General Accounting Office, *More Effective Use of Limited Training Funds*, 14.
28. William V. Kennedy, "For Cheney, Is 'Reserve' the Magic Word?" *Christian Science Monitor*, April 4, 1989, p. 19.
29. It is, of course, possible for a Guard officer commissioned through a state officer candidate school program to serve subsequently in another component.
30. Office of the Assistant Secretary of Defense (Reserve Affairs), *Official Guard and Reserve Manpower Strength and Statistics* (Washington, D.C., 1992), 2.180 (hereafter cited as *Guard and Reserve Manpower Strengths*).
31. *1992 Reserve Forces Policy Board Report*, 51.

32. AGR soldiers serve periodic tours and must be affirmatively retained in the program at the completion of each tour. AGR soldiers can voluntarily revert to part-time status and can serve nonrepetitive tours, that is, serve two AGR tours separated by part-time service. Thus, AGR soldiers who serve separate tours, as well as those who enter the AGR program as senior officers, can command as part-timers. Most AGR officers, however, regard the AGR program as a career and would not leave the program expressly for the opportunity to command as a part-timer.

33. Captain Richard J. Hyde, "A New Force Structure," *Military Review*, Nov. 1990, 16.

34. GAO *Readiness Report*, 17.

35. Colonel Michael A. Anastasio, "Leader Development: Direction for the Future," *Military Review*, May 1990, 12.

36. *1991 Reserve Forces Policy Board Report*, 41.

37. Barton Gellman, "Cheney Says Guard Units May Need Reorganizing," *Washington Post*, March 15, 1991, p. A34.

38. James W. Browning II et al., "The U.S. Reserve System: Attitudes, Perception, Realities," in Wilson, ed., *The Guard and Reserve in the Total Force*, 86.

39. Colonel David E. Shaver, *Closing Ranks: The Secret of Army Active and Reserve Component Harmony* (Carlisle Barracks, Pa.: U.S. Army War College Strategic Studies Institute, 1992), 3.

40. Lieutenant General Frederic J. Brown, "Reserve Forces: Army Challenge of the 1990s," *Military Review*, Aug. 1991, 10. See also Ritterpusch, *The Army Reserve*, 14: "On balance, . . . the Army's leaders are trained almost exclusively to perform functions required by the active component. Very little attention is paid to educating them about the Reserve environment or Reserve-unique problems."

41. House Committee on Appropriations, *Department of Defense Appropriations Bill, 1990: Report of the Committee on Appropriations*, 101st Cong., 1st sess., 1989, H. Rept. 208, 26.

42. Binkin and Kaufmann, *U.S. Army Guard and Reserve*, 59.

5. Lessons Learned from Desert Storm

1. Jim Tice, "Bush Calls Up IRR, Limits Reservists' Active Tours to One Year," *Army Times*, Feb. 4, 1991, p. 3.

2. Ibid.

3. Van D. Hipp, Jr., speech delivered at Columbia, S.C., July 8, 1991. Hipp is a former deputy assistant secretary of the Army for reserve affairs and mobilization.

4. Ibid.

5. William Matthews, "Call-up Bashes Political Myth," *Army Times*, Jan. 7, 1991, p. 9.

6. Steve Gibson, "Guards Flunked Fitness," *Sacramento Bee,* June 18, 1991, p. B1.
7. Lieutenant General James W. Crysel, speech delivered at Columbia, S.C., Jan. 1991.
8. Binkin and Kaufmann, *U.S. Army Guard and Reserve,* 93, quoting the Congressional Budget Office.
9. "Reserve Units Called Up," *Army Times,* Sept. 10, 1990, p. 18.
10. General Accounting Office, *Operation Desert Storm: Army Had Difficulty Providing Adequate Active and Reserve Support Forces* (Washington, D.C., 1992), 22-23 (hereafter cited as GAO *Support Report*).
11. Remarks of a reserve component brigadier general at a conference attended by the author, Jan. 19-20, 1991.
12. GAO *Support Report,* 25.
13. Testimony of General Colin L. Powell, chairman of the Joint Chiefs of Staff, before the Senate Armed Services Committee, Sept. 11, 1990 (recorded on C-SPAN); see also W. Matthews, "Call-up Bashes Political Myth," 8.
14. *Defense Authorization Act for Fiscal Year 1991,* Pub. L. 101-510, sec. 1112, 104 Stat. 1636 (1990).
15. Goldich, *Persian Gulf War,* 11.
16. Gellman, "Cheney Says," A34.
17. Richard Davis, *Operation Desert Storm: Army Guard Combat Brigade War Lessons Learned Reflect Long-Standing Problems* (Washington, D.C., 1992), 7 (text of statement delivered to House Armed Services Committee). Davis was formerly the director for Army issues of the General Accounting Office's National Security and International Affairs Division.
18. Ibid., 3.
19. Command Sergeant Major Richard Detjen, "AWOL and the Chain of Command," *Army,* April 1991, 6.
20. J. Paul Scicchitano, "Total Force or Total Failure?" *Army Times,* April 15, 1991, p. 14.
21. James C. Hyde, "Congress Skeptical that DoD Lives Up to Total Force Policy," *Armed Forces Journal International,* May 1991, 8.
22. Ibid.
23. General Robert W. Sennewald, "Fine Tuning Reserve Forces in the Wake of Desert Storm," *Army,* June 1991, 16.
24. GAO *Readiness Report,* 27.
25. J. Hyde, "Congress Skeptical," 8.
26. Andrew Shain, "Unit's War Readiness in Dispute," *[Columbia, S.C.] State,* Feb. 15, 1991, p. 3A.
27. "Were Reservists 'Used' to Save Army Budget?" *[Columbia, S.C.] State,* March 22, 1991, p. 10A.
28. Gellman, "Cheney Says," A34.
29. "The 48th Brigade: A Chronology from Invasion to Demobilization," *National Guard,* May 1991, 14.

30. See Jeffrey A. Jacobs, "A Mobilization Force," *Military Review*, Nov. 1990, 3-10.
31. Sennewald, "Fine Tuning Reserve Forces," 16.

6. Realizing the Potential of the Reserve Components

1. Testimony of General John W. Vessey, U.S. Army Retired, former chairman of the Joint Chiefs of Staff, before the Senate Armed Services Committee, Feb. 2, 1990 (recorded on C-SPAN).
2. Binkin and Kaufmann, *U.S. Army Guard and Reserve*, 97.
3. Lacy, "Whither the All-Volunteer Force?" 59.
4. Ibid.
5. Ibid., 67-68.
6. Ibid., 68.
7. Kennedy, "Is 'Reserve' the Magic Word?" 19.
8. J. Hyde, "Congress Skeptical," 8.
9. House Committee on Appropriations, *Department of Defense Appropriations Bill, 1990*, 24.
10. Binkin and Kaufmann, *U.S. Army Guard and Reserve*, 99-100. Additionally, the air reserve components have a much higher percentage of full-time support personnel.
11. W. Stanford Smith, "Reserve Readiness: Proving the Total-Force Policy a Success," in Wilson, ed., *The Guard and Reserve in the Total Force*, 118.
12. Colonel James L. Gould, "The Guard and Reserve: Towards Fuller Realization of the Total Force Potential," in *Mobilization and the National Defense*, ed. Hardy L. Merritt and Luther F. Carter (Washington, D.C.: National Defense Univ. Press, 1985), 125.
13. Fish, "The Constitution and the Training of National Guardsmen," 625, citing the congressional testimony of Assistant Secretary of Defense (Reserve Affairs) James H. Webb, Jr.
14. *Gray Board Report*, 14.
15. For my analysis of the constitutionality of federalizing the Guard, see "Reform of the National Guard: A Proposal to Strengthen the National Defense," *Georgetown Law Journal* 78 (Feb. 1990): 639-46.
16. Remarks of an ARCOM commander at a conference attended by the author, Jan. 19-20, 1991.
17. Field Manual 25-100, 1-9.
18. Binkin and Kaufmann, *U.S. Army Guard and Reserve*, 104, citing a U.S. Army Training Board study.
19. General Accounting Office, *More Effective Use of Limited Training Funds*, 13.
20. This general concept was proposed by Major General John L. Matthews, adjutant general of Utah, in "Assessing Reserve Component Training," *Military Review*, Nov. 1989, 27-33.

21. The Reserve Officer Personnel Management Act would institute a below-the-zone system.
22. Binkin and Kaufmann, *U.S. Army Guard and Reserve*, 72.
23. See also R. Hyde, "A New Force Structure," 17.
24. See U.S. House of Representatives, *Defense Authorization Act for Fiscal Year 1991*, Report of the Conference Committee, 101st Cong., 2d sess., 1990, H. Conf. Rept. 923, 600.
25. Kennedy, "The Reserve Dilemma," A24.
26. J. Matthews, "Assessing Reserve Component Training," 15.
27. WESTAT, Inc., *Survey*, 437-38; see also White and Simpson, "Why Soldiers 'Really' Leave," 76.
28. J. Hyde, "Congress Skeptical," 8.
29. *1989 Reserve Forces Policy Board Report*, 41.
30. *Partners in the Total Army*, 10
31. J. Matthews, "Assessing Reserve Component Training," 31-32.
32. Lewis Sorley, "Creighton Abrams and Active-Reserve Integration in Wartime," *Parameters* 21 (Summer 1991): 44-46.
33. See the ideas of Senator Sam Nunn (D-Georgia), chairman of the Senate Armed Services Committee, in Soraya S. Nelson, "Nunn: Rewrite Strategy, Rely on Reserves," *Army Times*, April 30, 1990, p. 19.
34. Goldich, *Persian Gulf War*, 11. Notably, however, the Marine Corps maintains a much more integrated active-reserve structure and provides much more full-time active component support to its reserves than does the Army.
35. Brown, "Reserve Forces," 8.
36. Tom Donnelly, "Sullivan Putting New Spin on Roundout Brigades," *Army Times*, Sept. 16, 1991, p. 10.
37. Lieutenant General Herbert R. Temple, Jr., "Training RC Soldiers," *Army*, Jan. 1990, 2.
38. House Committee on Armed Services, *Defense Authorization Act for Fiscal Year 1991*, 101st Cong., 2d sess., 1990, H. Rept. 665, 268-69; General Hamilton H. Howze, "Shrink Army if We Must, But Don't Hollow it Out," *Army*, May 1990, 11, and "Toward Real Reserve Readiness: The Case for the Cadre System," *Army*, Aug. 1972, 12.
39. See Lieutenant Colonel Thomas R. Rozman and Lieutenant Colonel William A. Saunders, "The Expansible Army," *Military Review*, Nov. 1990, 37-38.
40. Department of Defense, *Total Force Policy Report to the Congress* (Washington, D.C., 1990), 64.
41. Colonel Charles E. Heller, *The New Military Strategy and Its Impact on the Reserve Components* (Carlisle Barracks, Pa.: U.S. Army War College Strategic Studies Institute, 1991), 26.

7. Through the Political Minefield

1. Lacy, "Whither the All-Volunteer Force?" 68.
2. Ibid., citing Martha Derthick, "Militia Lobby in the Missile Age,"

in *Changing Patterns of Military Politics,* ed. Samuel Huntington (New York: Free Press of Glencoe, 1962).

3. Rick Maze, "Group Wants Independent Reserves Study," *Army Times,* July 1, 1991, 4.

4. Dave Moniz, "Guard Fighting for Life," *[Columbia, S.C.] State,* April 5, 1992, p. 1B.

5. See Dennis Steele, "150th—A Number on a List in DC; Gut Blow in Welch," *Army,* May 1990, 30-39.

6. House Committee on Armed Services, *National Defense Authorization Act for Fiscal Year 1993: Report to Accompany H.R. 5006,* 102d Cong., 2d sess., 1992, H. Rept. 527, 398-99.

7. Ritterpusch, *The Army Reserve,* 34.

8. "A Force of 'Great Utility' That Cannot be Mass-Produced," *Army,* April, 1992, 30.

Appendix

1. 10 U.S.C. § 162-63 (1988).
2. 10 U.S.C. § 151-53, 163 (1988).
3. Office of the Joint Chiefs of Staff, Joint Publication 1-02, *Department of Defense Dictionary of Military and Associated Terms* (Washington, D.C., 1989), 75-76.
4. 10 U.S.C. § 261 (1988).
5. 10 U.S.C. § 175 (1988).
6. *Training and Organization of the Reserve Components,* 15.
7. 10 U.S.C. § 3021 (1988).
8. Binkin and Kaufmann, *U.S. Army Guard and Reserve,* 28.
9. *1992 Reserve Forces Policy Board Report,* 13
10. *Guard and Reserve Manpower Strengths,* 1.005.
11. *1992 Reserve Forces Policy Board Report,* 51.
12. Ibid.
13. *Guard and Reserve Manpower Strengths,* 1.005.
14. *1991 Reserve Forces Policy Board Report,* 11.
15. *1990 Reserve Forces Policy Board Report,* 25, 29.
16. Ibid.
17. *Guard and Reserve Manpower Strengths,* 1.005.
18. Ibid.
19. Ibid.
20. 10 U.S.C. § 688 (1988).
21. 10 U.S.C. § 311 (1988).

SELECTED BIBLIOGRAPHY

"A Force of 'Great Utility' That Cannot be Mass-Produced." *Army*, April 1992, 24-33.
Anastasio, Michael A. "Leader Development: Direction for the Future." *Military Review*, May 1990, 10-19.
Association of the United States Army. *The Active and Reserve Components: Partners in the Total Army.* Arlington, Va., 1989.
———. *Divisions of the U.S. Army.* Arlington, Va., 1989. (Foldout wall chart.)
———. *The Future U.S. Army—Outlook for the 1990s.* Background Brief No. 12. Arlington, Va., 1990.
Binder, L. James. "Vital to Army that Reductions be Managed." *Army*, Feb. 1990, 14-18.
Binkin, Martin, and William W. Kaufmann. *U.S. Army Guard and Reserve: Rhetoric, Realities, Risks.* Washington, D.C.: Brookings Institution, 1989.
Brehm, Philip A. *Restructuring the Army: The Road to a Total Force.* Carlisle Barracks, Pa.: U.S. Army War College Strategic Studies Institute, 1992.
Brown, Frederic J. "Reserve Forces: Army Challenge of the 1990s." *Military Review*, Aug. 1991, 3-19.
Browning, James W., II, Kenneth C. Carlon, Robert L. Goldich, Neal F. Herbert, Theodore R. Mosch, Gordon R. Perkins, and Gerald W. Swartzbaugh. "The U.S. Reserve System: Attitudes, Perception, Realities." In *The Guard and Reserve in the Total Force,* edited by Bennie J. Wilson III. Washington, D.C.: National Defense Univ. Press, 1985.
Burdick, Donald. "Army National Guard: An Essential Element of National Strategy." *Army*, Oct. 1990, 116-21.
Carlton, John T., and John F. Slinkman. *The ROA Story.* Washington, D.C.: Reserve Officers Association, 1982.
Committee on Civilian Components. *Reserve Forces for National Security: Report to the Secretary of Defense.* Washington, D.C., 1948.
Crossland, Richard B., and James T. Currie. *Twice the Citizen: A History of*

the Army Reserve, 1908-1983. Washington, D.C.: Office of the Chief, Army Reserve, 1984.

Davis, Richard. *Operation Desert Storm: Army Guard Combat Brigade War Lessons Learned Reflect Long-Standing Problems.* Washington, D.C., 1992. Text of statement delivered to House Armed Services Committee.

Defense Authorization Act for Fiscal Year 1991, Pub. L. 101-510, 104 Stat. 1620 (1990).

Defense Manpower Data Center. *Descriptions of Officers and Enlisted Personnel in the U.S. Selected Reserve.* Arlington, Va., 1987.

Department of the Army. CAPSTONE *Program.* Army Regulation 11-30. Washington, D.C., 1985.

———. *Operations.* Field Manual 100-5. Washington, D.C., 1986.

———. *Training the Force.* Field Manual 25-100. Washington, D.C., 1988.

Detjen, Richard. "AWOL and the Chain of Command." *Army,* April 1991, 6-7.

Department of Defense. *Total Force Policy Report to the Congress.* Washington, D.C., 1990.

Fish, Peter A. "The Constitution and the Training of National Guardsmen: Can State Governors Prevent Uncle Sam from Sending the Guard to Central America?" *Journal of Law and Politics* 4 (Winter 1988): 597-637.

"The 48th Brigade: A Chronology from Invasion to Demobilization." *National Guard,* May 1991, 12-15.

Galtney, Liz. "The Sad State of Weekend Warriors." *U.S. News & World Report,* Sept. 25, 1989, 28-29.

General Accounting Office. *Army Reserve Components: Minimum Essential Equipment for Training Has Not Been Effectively Managed.* Washington, D.C., 1990.

———. *Army Training: Evaluations of Units' Proficiency Are Not Always Reliable.* Washington, D.C., 1991.

———. *Army Training: Management Initiatives Needed to Enhance Reservists' Training.* Washington, D.C., 1989.

———. *Desert Shield/Storm Logistics: Observations by U.S. Military Personnel.* Washington, D.C., 1991.

———. *Individual Ready Reserve: Army Needs to Make More Effective Use of Limited Training Funds.* Washington, D.C., 1990.

———. *National Guard: Peacetime Training Did Not Adequately Prepare Combat Brigades for Gulf War.* Washington, D.C., 1991.

———. *Operation Desert Storm: Army Had Difficulty Providing Adequate Active and Reserve Support Forces.* Washington, D.C., 1992.

———. *Reserve Components: Opportunities to Improve National Guard and Reserve Policies and Programs.* Washington, D.C., 1988.

Goldich, Robert L. *The Army's Roundout Concept after the Persian Gulf War.* Washington, D.C.: Congressional Research Service, 1991.

———. *Persian Gulf War: U.S. Reserve Callup and Reliance on the Reserves.* Washington, D.C.: Congressional Research Service, 1991.

SELECTED BIBLIOGRAPHY

———. *U.S. Army Combat-to-Support Ratios: A Framework for Analysis.* Washington, D.C.: Congressional Research Service, 1989.

Gordon, Sol, ed. *1986 National Guard Almanac.* Washington, D.C.: Uniformed Services Almanac, 1986.

Gould, James L. "The Guard and Reserve: Towards Fuller Realization of the Total Force Potential." In *Mobilization and the National Defense,* edited by Hardy L. Merritt and Luther F. Carter. Washington, D.C.: National Defense Univ. Press, 1985.

Heller, Charles E. *Economy of Force: A Total Army, the Israel Defense Force Model.* Carlisle Barracks, Pa.: U.S. Army War College Strategic Studies Institute, 1992.

———. *The New Military Strategy and Its Impact on the Reserve Components.* Carlisle Barracks, Pa.: U.S. Army War College Strategic Studies Institute, 1991.

Hill, Jim Dan. *The Minuteman in Peace and War.* Harrisburg, Pa.: Stackpole, 1964.

Hirsch, Alan. "The Militia Clauses of the Constitution and the National Guard." *University of Cincinnati Law Review* 56 (1988): 919-69.

Howze, Hamilton H. "Shrink Army if We Must, But Don't Hollow it Out." *Army,* May 1990, 11-14.

———. "Toward Real Reserve Readiness: The Case for the Cadre System." *Army,* Aug. 1972, 12-16.

Hyde, James C. "Congress Skeptical that DoD Lives Up to Total Force Policy." *Armed Forces Journal International,* May 1991, 8.

Hyde, Richard J. "A New Force Structure." *Military Review,* Nov. 1990, 11-18.

Jacobs, Jeffrey A. "Integrating the Total Army: The Road to Reserve Readiness." *Parameters* 20 (Dec. 1990): 73-84.

———. "A Mobilization Force." *Military Review,* Nov. 1990, 3-10.

———. "RC Individual Training Suffers During Higher Level Emphasis." *Army,* Nov. 1989, 14-15.

———. "Reform of the National Guard: A Proposal to Strengthen the National Defense." *Georgetown Law Journal* 78 (Feb. 1990): 625-47.

———. "Today's National Guard: Time to Federalize." *Armed Forces Journal International,* April 1990, 52-58.

Lacy, James L. "Whither the All-Volunteer Force?" *Yale Law & Policy Review* 5 (Fall/Winter 1986): 38-72.

Mahon, John K. *History of the Militia and the National Guard.* New York: Macmillan, 1983.

Marshall, S.L.A. "McNamara's Latest Reform: Why His National Guard Scheme Won't Work." *New Republic,* Jan. 23, 1965, 13-15.

Matthews, John L. "Assessing Reserve Component Training." *Military Review,* Nov. 1989, 27-33.

Moxon, Arthur L. "U.S. Reserve Forces: The Achilles' Heel of the All-Volunteer Force?" In *The Guard and Reserve in the Total Force,* edited by Bennie J. Wilson III, 91-113. Washington, D.C.: National Defense Univ. Press, 1985.

Newland, Samuel J. "The National Guard: Whose Guard Anyway?" *Parameters* 18 (June 1988): 40-50.
Office of the Assistant Secretary of Defense (Reserve Affairs). *Official Guard and Reserve Manpower Strength and Statistics.* Washington, D.C., 1992.
Office of the Chief, Army Reserve. *The Chief, Army Reserve's Long-Range Plan, 1990-2020.* Washington, D.C., 1990.
Office of the Secretary of Defense. *Reserve Component Programs—FY 1992: Report of the Reserve Forces Policy Board.* Washington, D.C., 1993.
———. *Reserve Component Programs, Fiscal Year 1991: Report of the Reserve Forces Policy Board.* Washington, D.C., 1992.
———. *Reserve Component Programs, Fiscal Year 1990: Report of the Reserve Forces Policy Board.* Washington, D.C., 1991.
———. *Reserve Component Programs: Report of the Reserve Forces Policy Board, Fiscal Year 1989.* Washington, D.C., 1990.
———. *Reserve Component Programs, Fiscal Year 1988: Report of the Reserve Forces Policy Board.* Washington, D.C., 1989.
Palmer, Bruce, Jr. *The 25-Year War: America's Military Role in Vietnam.* Lexington: Univ. Press of Kentucky, 1984.
Philbin, Edward J., and James L. Gould. "The Guard and Reserve: In Pursuit of Full Integration." In *The Guard and Reserve in the Total Force,* edited by Bennie J. Wilson III, 43-57. Washington, D.C.: National Defense Univ. Press, 1985.
Powell, Colin L. *National Military Strategy of the United States.* Washington, D.C., 1992.
Rees, Raymond F. "The Largest Mobilization in 30 Years." *Army,* October 1991, 110-13.
Ritterpusch, David S. "The Army Reserve and U.S. National Defense Strategy." U.S. Army War College Military Studies Program Paper, 1988.
Rozman, Thomas R., and William A. Saunders. "The Expansible Army." *Military Review,* Nov. 1990, 30-39.
Sandler, Roger W. "When the Call Came Reserve Was Ready." *Army,* Oct. 1991, 100-106.
Sennewald, Robert W. "Fine Tuning Reserve Forces in the Wake of Desert Storm." *Army,* June 1991, 14-17.
Shaver, David E. *Closing Ranks: The Secret of Army Active and Reserve Component Harmony.* Carlisle Barracks, Pa.: U.S. Army War College Strategic Studies Institute, 1992.
Smith, W. Stanford. "Reserve Readiness: Proving the Total-Force Policy a Success." In *The Guard and Reserve in the Total Force,* edited by Bennie J. Wilson III, 115-122. Washington, D.C.: National Defense Univ. Press, 1985.
Sorley, Lewis. "Creighton Abrams and Active-Reserve Integration in Wartime." *Parameters* 21 (Summer 1991): 35-50.
Temple, Herbert R., Jr. "Training RC Soldiers." *Army,* Jan. 1990, 2.

SELECTED BIBLIOGRAPHY

Upton, Emory. *The Military Policy of the United States*. Washington, D.C., 1907.
U.S. Army Training Board. *Training and Organization of the Reserve Components: A Reference Text for Total Force Trainers, 1987-88*. Ft. Monroe, Va., 1988.
U.S. Congress. House. *Defense Authorization Act for Fiscal Year 1991*. Report of the Committee on Armed Services. 101st Cong., 2d sess., 1990, H. Rept. 665.
―――. *Defense Authorization Act for Fiscal Year 1991, Report of the Conference Committee*. 101st Cong., 2d sess., 1990, H. Conf. Rept. 923.
―――. *Department of Defense Appropriations Bill, 1990, Report of the Committee on Appropriations*. 101st Cong., 1st sess., 1989, H. Rept. 208.
―――. *National Defense Authorization Act for Fiscal Year 1993, Report of the Committee on Armed Services*. 102d Cong., 1st sess., 1992, H. Rept. 527.
Velez, German J. "48 Hours: Fighting the Reserve Component Battle." *Infantry*, Nov.-Dec. 1989, 16-17.
Vuono, Carl E. "Change, Continuity and the Army of the 1990s," *Army*, June 1991, 9-12.
Weigley, Russell F. *History of the United States Army*. New York: Macmillan, 1967.
WESTAT, Inc. *Survey of United States Army Reserve (USAR) Troop Program Units (TPU) Soldiers*. Rockville, Md., 1988.
White, Terry, and Jim Simpson. "Why Soldiers 'Really' Leave the Army Reserve: The WESTAT Survey." *Military Review*, Feb. 1990, 76-77.
Wickham, John A., Jr. *Chief of Staff's White Paper on the Light Infantry Division, Army of Excellence*. Washington, D.C., 1984.
Wilson, Bennie J., III, and James R. Engelage. "Pretrained Individual Manpower: Albatross or Phoenix?" In *The Guard and Reserve in the Total Force*, edited by Bennie J. Wilson III, 123-146. Washington, D.C.: National Defense Univ. Press, 1985.
Wiener, Frederick B. "The Militia Clauses of the Constitution." *Harvard Law Review* 54 (Dec. 1940): 181-220.

INDEX

Abrams, Creighton, 130
active component, 5-7; assignment and promotion policies, 53, 86, 139; evaluation of reserves, 63-64; focus of Army on, 53. *See also* CAPSTONE: active Army perspective; cultural difference between active and reserve components; United States Forces Command: active component
Adams, John, 29
Adjutant Generals' Association, 39, 137
adjutants general, 9-10, 18, 55, 57-60, 100
administrative requirements, 55
affiliation relationship, 17-18
AGR (active Guard/Reserve) personnel. *See* full-time support personnel: AGR (active Guard/Reserve)
Air National Guard, 9, 10-11, 24, 42, 69, 106-07, 111, 151
annual training, 8, 21, 37, 57-59, 70; evaluation of, 18, 63, 143; in proposed system, 126-28
ARCOMs (Army Reserve commands), 14, 18, 24, 61-62, 111-12, 114-15
Armed Forces Reserve Act of 1952, 43, 67, 104
Army National Guard: in civil rights crises, 44; dual status, 9, 39-40; funding of, 10-11, 36-37, 55-56, 59, 109; organization, 9-11; state commissioning programs, 79, 111; state control of, 9-10, 44, 55-60, 108-11. *See also* militia

Army National Guard of the United States, 9, 24, 39-40, 67
Army National Guard Personnel Center, 21, 111, 121, 129
Army Reserve. *See* United States Army Reserve
Army Reserve commands. *See* ARCOMs
Army Reserve Personnel Center, 77, 121, 129
Army Training and Evaluation Program, 112-13
Articles of Confederation, 27-28
Aspin, Les, 99
assistant secretary of defense (reserve affairs), 54, 98, 151
assistant secretary of the Army (manpower and reserve affairs), 151

battle focus, 113
Berlin crisis, 44
Binkin, Martin, 26, 68, 86, 107
bonuses, enlistment and reenlistment, 128
Burba, Edwin H., 98
Bush, George, 88

cadre divisions, 133
Calhoun, John C., 30, 32, 42
CAPSTONE, 16-19, 114; active Army perspective, 62-63; in Desert Storm, 92-95; funds, 18; guidance from wartime headquarters, 54, 59, 112-13; practical effect of, 62-64
chairman, Joint Chiefs of Staff, 89, 150

Cheney, Dick, 98, 100
chief, Army Reserve, 11, 39. *See also* office of the chief, Army Reserve
chief, National Guard Bureau, 11, 59, 133
chief of staff of the Army, 11, 53, 132, 150
Civil War, United States, 31, 33
cohesiveness of units. *See* personnel turnover
combatant commands, 61, 87, 118, 147-50. *See also* specified commands; unified commands
Combat Maneuver Training Center, 52
combat support and combat service support units, 11, 12, 101, 125, 151; in contingency role, 130, 131, 134-35; in Desert Storm, 89-90, 101, 105, 130; pay disparity in proposed system, 128; as percentage of components, 71, 74; readiness of, 87, 101, 124-25; training, 101
combat training centers, 52
combat units, 11, 12, 151; contingency missions, 130-31; effect of geography and time on, 69, 125; importance of, in reserve components, 99-100, 132; training, 101-02, 130-32
command selection, 121-22
common task test, 71
Constitution, United States: army clause, 9, 39-40, 109-10; militia clauses, 9, 28-29, 34-46, 39-40, 56
continental United States armies. *See* CONUSAs
contingency missions, 5, 130-31, 134-35
CONUSAs (continental United States armies), 14-16; relationship to Army National Guard units, 57-59; streamlining of, 114, 129; transition to USARC, 61-62
cross-leveling, 118
Crysel, James W., 98
cultural difference between active and reserve components, 85-86, 94, 124, 170 n 40

Dearborn, Henry, 29
demographics, effect of, 72, 74
deputy chief of staff for logistics, 53
deputy chief of staff for operations and plans, 53
deputy chief of staff for personnel, 53
Desert Shield/Desert Storm, 2, 4, 7, 48, 88-89; age of reservists, 85; CAPSTONE, 92-95, 114; combat support and combat service support units in, 89-90, 101, 105, 130; combat units, ineffectiveness of, 101-02; Individual Ready Reserve in, 89, 102; maintenance, 96; mobilization for, 88-89; politics surrounding mobilization decisions, 96, 100-101; postmobilization training, 90, 95-97, 102, 129; public support for, 134-35; readiness reporting, 90-91, 115; roundout, 89, 92, 95-101, 131-32; Total Force Policy in, 2, 52, 88, 102
Dick, Charles, 34
Dick Act, 34-35
directed training association, 16-17, 63-64
director, Army National Guard, 11
dispersion of reserve units, 72, 124
Downing, Wayne A., 144
drills. *See* unit training assemblies
Duncan, Stephen M., 98

education. *See* professional development
Eighth Army, 117
8th Infantry Division (Mechanized), 163 n 3
82d Airborne Division, 5
Eisenhower, Dwight D., 44
Enlisted Reserve Corps, 37, 38, 40
equipment incompatibility, 51-52

Faubus, Orville, 44
federal recognition, 9-10, 77
5th Infantry Division (Mechanized), 56, 96, 97
1st Cavalry Division, 56, 95
1st Infantry Division (Mechanized), 56

INDEX

Forces Command. *See* United States Forces Command
Forrestal, James, 42
48th Infantry Brigade (Mechanized), 85, 95-97
4th Infantry Division (Mechanized), 56
Full-Time Support Management Center, 21, 80
full-time support personnel, 21-22, 79-81, 82, 122; AGR (active Guard/Reserve), 21-22, 80-81, 111, 122-24, 143, 155, 170 n 32; military technicians, 22, 122-23, 155
funding of reserves, 22-23, 51, 129
funds, training, 114

Garrison, Lindley, 36, 37
general officer commands, 14, 61
geography, effect of, 68-70, 72-75, 77, 78, 124-25
Gould, James L., 108
Gray, Gordon, 42
Gray Board, 42, 45, 50, 59, 60, 109, 141
Grenada operation, 8

Hay, James, 36
Heller, Charles E., 134
Home Guard. *See* State Guard

Inactive National Guard, 155
incompatible supply and personnel systems, 53. *See also* equipment incompatibility
individual mobilization augmentees, 8, 155
Individual Ready Reserve, 8-9, 64-67, 76, 129, 155, 157; in Desert Storm, 89, 102; mobilization of, generally, 19; proposed reforms of, 118-19; screen, 65, 119
Inouye, Daniel K., 106
Israel, reserves, 106

Joint Readiness Training Center, 52

Kaufmann, William W., 26, 68, 86, 107
Kennedy, John F., 44

Kennedy, William V., 78, 105, 126, 136
Korean War, 42-43

Lacy, James L., 54, 64, 104-05, 136
Lincoln, Abraham, 31
litigation over deployment of National Guard units, 47-48, 56

McNair, Lesley, 41
McNamara, Robert S., 45, 138, 140
Mahon, John K., 29
maintenance, in Desert Storm, 96
March, Peyton C., 38
Marsh, John O., Jr., 138
Medical Reserve Corps, 35, 37
Mexican War, 30-31, 33
military technicians, 22, 122-23, 155
militia, 9, 158; foreign service, 29, 33, 35, 37, 40, 108; history of, 27-37, 40; National Guard as, 9, 40, 56, 108-09, 151; state control of, 34-35
Militia Act of 1792, 28, 33, 39
Militia Bureau, 37, 39
Minimum Essential Equipment for Training, 51
mission essential task list, 54, 84, 113
mobilization, 8, 10, 19-20; for Desert Storm, 88-89; for Korean War, 42-43; for Vietnam War, 45-46; for World War I, 37-38; for World War II, 40-41
Monroe, James, 30

National Defense Act of 1916, 36-37, 40
National Defense Act of 1933, 39-40, 67
National Guard. *See* Air National Guard; Army National Guard
National Guard Association, 31-32, 34, 39, 41; and Desert Storm, 100; influence on Congress, 37, 41-42, 136; opposition to reforms, 36, 38-39, 42-44
National Guard Bureau, 10-11, 21, 24, 56, 59, 111
National Guard of the United

States. *See* Army National Guard of the United States
national manpower pool system, 125-29
National Training Center, 52-53, 96-97
9th Infantry Division (Motorized), 163 n 3

office of the chief, Army Reserve, 11, 21
Officers' Reserve Corps, 37, 38, 40, 41
155th Armored Brigade, 95-96
101st Airborne Division (Air Assault), 5
197th Infantry Brigade, 95
Operation Just Cause, 8
Organized Reserve Corps, 38, 40, 41

Pagonis, William G., 93
Palmer, Bruce, 46
Palmer, John McAuley, 1, 38, 88, 138
pay for reserve component personnel, 23, 128, 157
Pershing, John J., 38
personnel turnover, 72-73
physical fitness, 84-85, 89-90
Pike, Zebulon, 30
politics: and Desert Storm, 96, 100; intra-unit, 35, 48, 78, 111, 121, 125 and reserves, 26, 29, 31, 33-34, 48, 83, 136-42, 145
Polk, James K., 31
postmobilization training. *See* training, postmobilization
Powell, Colin L., 3-4, 89, 95
president: constitutional role, 147; role in mobilization, 19, 43
professional development, 81-84, 119-20
promotion policies, 75-78, 121-22. *See also* active component: assignment and promotion policies

readiness groups, 14-15, 86, 114, 129
readiness regions, 86
readiness reporting, 64, 90-91, 114

Ready Reserve, 8-9, 19, 20, 43, 67, 155, 157
recruiting in reserve units, 72, 78-79
regional training centers, 69
Reserve Component Automation System, 167 n 7
Reserve Forces Policy Board, 55, 71-72, 75, 151
Reserve Forces Policy Committee, 153
Reserve Officer Personnel Management Act, 76, 173 n 21
Reserve Officers Association, 39, 44, 45, 136, 137
Reserve Officers' Training Corps (ROTC), 37, 39, 79
Retired Reserve, 43, 157-58
retraining of reserve personnel, 71, 75
rivalry between reserve components, 43, 53-54, 111
Roosevelt, Theodore, 104
Root, Elihu, 33-35, 37
roundout, 16-17, 52, 56; in Desert Storm, 89, 92, 95-101; elimination of, 131-32; readiness of roundout units, 50
roundup, 56

Scott, Winfield, 30, 31
2d Armored Division, 95, 163 n 3
2d Infantry Division, 56
secretary of defense, 19, 138, 147, 150, 151
secretary of the Army, 11, 138, 150
Selected Reserve: mobilization of, 19-20, 88; organization, 8, 54; skill qualification of soldiers, 71, 75; training, 21, 105; use as fillers, 118
senior leadership, 82, 83-84
Sennewald, Robert W., 99, 101
7th Infantry Division (Light), 5
Shaver, David E., 86
6th Infantry Division (Light), 17, 56
skill qualification test, 71, 169 n 12
Soviet Union, 1, 4
Spanish-American War, 33
specified commands, 7, 50, 147,

INDEX

148-49. *See also* combatant commands
Standby Reserve, 43, 155, 157
state area commands, 10, 60, 111
State Guard, 37, 41
Stimson, Henry L., 36
Stone, Michael P.W., 74
Strategic Air Command, 147
Switzerland, reserves, 106

Taylor, Maxwell D., 44
Taylor, Zachary, 31
Temple, Herbert R., Jr., 51
tension between reserves and Regular Army, 29-31, 33, 38, 40, 42, 44, 48, 100
10th Mountain Division (Light Infantry), 56
3d Armored Division, 163 n 3
Third Army, 92, 115
35th Infantry Division (Mechanized), 57
threats to United States interests, 3-4
III Corps, 112, 117
335th Signal Command, 92-93
377th Theater Army Area Command, 92-93
time, effect of, 69-70, 125
Total Army Personnel Command, 121
Total Force Policy, 2-3, 47, 135, 145; affecting readiness of Army National Guard, 109; Army support of, 86, 137-38; in Desert Storm, 88, 89, 102; increased importance of, 4; integration of components, 23-24, 54, 93-94, 99, 104, 105; role of active component in, 14-16; role of reserves in, 7-8, 47, 50, 129-30, 133-35, 153-55
Total Force Policy Report, 134
training, individual, 70-72
training, postmobilization, 20, 87, 98, 129, 130-33; in Desert Storm, 90, 95-97, 102; for Vietnam conflict, 45; in World War I, 37-38; in World War II, 41
training areas, 68-70, 124-25
training divisions, 133, 143
25th Infantry Division (Light), 5

24th Infantry Division (Mechanized), 56, 95
29th Infantry Division (Light), 10, 57-60
200K call-up. *See* mobilization
256th Infantry Brigade (Mechanized), 96, 97

unified commands, 50, 88, 112, 147, 148-149. *See also* combatant commands
unit conversions, 73-74
United States Air Force, 9, 10, 69, 106-07, 147
United States Air Force Reserve, 69, 106-07, 151
United States Army Civil Affairs and Psychological Operations Command, 144
United States Army Europe, 5, 11, 46, 112, 117
United States Army Information Systems Command, 93
United States Army Pacific, 7, 11
United States Army Recruiting Command, 143
United States Army Reserve: control by administrative headquarters, 61-62, 111-12; organization, 11-14; retraining of soldiers from active component, 71
United States Army Reserve Command. *See* USARC
United States Army Reserve Forces schools, 82, 120, 129
United States Army Special Operations Command, 11, 143-44
United States Army Training and Doctrine Command, 120
United States Central Command, 88, 90, 92, 93, 147
United States Coast Guard, 147
United States Coast Guard Reserve, 151
United States European Command, 112
United States Forces Command: active component units, 7, 14, 63, 112; command of Army Reserve units, 11, 60-61; deputy commander, 92; in Desert Storm, 93; reserve units, training of,

120; role in proposed system, 117
United States Marine Corps, 132, 147, 173 n 34
United States Marine Corps Reserve, 151, 173 n 34
United States Naval Reserve, 107, 151
United States Navy, 147
United States property and fiscal officers, 10, 111
United States Southern Command, 8
unit training assemblies, 8, 21, 37, 69-70, 126, 128
unit vacancy promotion system, 77-78, 122
universal military training, 39, 106
up-or-out system, 76-77, 119
Upton, Emory, 32-33, 35, 42, 138, 142
USARC (United States Army Reserve Command), 14, 16, 114; command relationships, 18, 24, 61-62, 143; relationship with CONUSAs, 15-16; role in CAPSTONE, 112

Vietnam War, 1-2, 45-46, 89, 130
Villa, Pancho, 36
Volunteers, 1, 30-31, 33

Wagner, Robert E., 138
Wallace, George, 44
War of 1812, 29-30
wartime gaining commands. *See* CAPSTONE
Webb, James H., Jr., 73
weekend drills. *See* unit training assemblies
Weigley, Russell, 32
Wiener, Frederick B., 40
Wilson, Woodrow, 36
Wood, Leonard, 35-36, 37
World War I, 1, 3, 37-38, 41
World War II, 1, 40-41

www.ingramcontent.com/pod-product-compliance
Lightning Source LLC
Chambersburg PA
CBHW032045150426
43194CB00006B/422